Student Study & Solutions Guide

Essentials of

General, Organic, and Biological Chemistry

Armold

Owen McDougal

SOUTHERN OREGON UNIVERSITY

Australia • Canada • Mexico • Singapore • Spain • United Kingdom • United States

Brooks/Cole — Thomson Learning
10 Davis Drive
Belmont CA 94002-3098
USA

For information about our products, contact us:
Thomson Learning Academic Resource Center
1-800-423-0563
http://www.brookscole.com

For permission to use material from this text, contact us by
Web: http://www.thomsonrights.com
Fax: 1-800-730-2215
Phone: 1-800-730-2214

Printed in the United States of America
2 3 4 5 6 7 09 08 07 06

PREFACE

This study guide is designed for the student that has both limited time and a limited understanding of chemistry. The aim of the study guide is to increase student comprehension of the material presented in Essentials of General, Organic, and Biological Chemistry by Mel Armold. Each chapter of the study guide contains chapter objectives, definitions to key terms in the order in which they appear in the text, detailed descriptions of the objectives, solutions as well as answers to odd numbered problems, a sample test, and answers to the sample test.

It should be noted that the majority of people who take chemistry do not understand everything that is presented to them. The study guide is designed to help you the student get the most out of your study time. There is a system for studying college chemistry that can be broken down by student type. The "A" student will:

a.) Read the text book before and after lecture
b.) Take notes in lecture
c.) Rewrite the lecture notes making sure to fill in the blanks on confusing topics
d.) Ask questions concerning difficult concepts
e.) Do all of the problems in the text prior to checking the answers in the back of the book
f.) Read the study guide and take the sample test
g.) Check the answers to the sample test

The reality is that very few students can afford to put in this kind of time and effort into one subject. For the students who want to get to the point and do well, the study guide is set up like a set of "Cliff Notes". The format is broken down into six main categories.

1.) **Chapter Objectives:** The chapter objectives are listed so the student knows the general areas they should be able to address. This is nothing more than an overview or one sentence description of what is discussed.

2.) **Key Terms:** The key terms for the chapter are defined in the order that they are encountered in the text. With the study of science comes a new vocabulary that is the basis for learning. Much like a foreign language if you don't know the words, you can't make sense of the sentence. The key term section is to be used as a quick reference as well as a clearly stated definition of important words.

3.) **Objective Details:** The chapter objectives are highlighted. It should not be assumed that all of the information in the text is available in the objective details, however the main points to each section of the textbook have been indicated. There is no substitute to reading the text. If there is something in the textbook that you do not understand, the topic is most likely addressed in the objective details in a slightly different way. This subtle difference in explanation can sometimes make a big difference in understanding.

4.) **Solutions:** The solutions to odd numbered textbook problems are provided to show the approach to solving the problems as well as to give the answer to the problem. In many aspects of chemistry, the majority of credit can be obtained if the process is correct even if the answer is incorrect.

5.) **Sample Test:** When you feel confident with your abilities to solve problems and have a good understanding of the material, you can test your abilities with the sample test. Take the entire test prior to looking at any of the answers. The questions are in order of sections in the textbook. There is at least one question for each objective in the chapter. Be honest to yourself and take the test as a test. It is better to miss questions here than to miss them when they count for credit.

6.) **Test Answers:** The answers to the sample test are provided to eliminate questions concerning correct answers. If you miss questions from the sample test, return to that area of the textbook and do more problems from that section.

Questions or corrections regarding this study guide should be emailed to mcdougao@sou.edu or sent to the following address:

Owen McDougal, PhD.
1250 Siskiyou Boulevard
Department of Chemistry
Southern Oregon University
Ashland, OR 97520

TABLE OF CONTENTS

Chapter 1: The Principles and Tools of Science

Chapter Objectives:

1. To understand how data, hypotheses, and experimentation contribute to the scientific method.
2. To gain a broad understanding of common units associated with the Systѐme International, metric system, and the English system of measurement.
3. To become familiar with unit prefixes.
4. To incorporate accuracy and precision into discussions of measurement.
5. To identify what values in a measurement are significant figures.
6. To use scientific notation to express the correct number of significant figures and exponential powers of ten.
7. To use conversions between units regardless of the system of measurement.
8. To understand matter and the three states of matter (solid, liquid, and gas).
9. To differentiate among elements, substances, compounds, and mixtures.
10. To understand the relationship between kinetic energy, potential energy, heat, and temperature.

Key Terms: The key terms are listed in the order in which they are encountered in the Chapter.

Scientific method. The process by which data is gathered, a hypothesis is made, and is then experimentally tested for validity.

Data. Specific factual information.

Hypothesis. A statement that explains a known set of data and makes predictions that can be tested by experimentation.

Theory. A credible and broadly accepted statement that emerges from one or more accepted hypotheses for the explanation of a set of data.

Law. A generalization or summary that is based on consistent experimental results or experiences.

Diagnosis. In medicine, a diagnosis is a hypothesis that predicts the best course of treatment for an ailment.

Measurement. Observations that provide numerical information about a subject or topic.

Quantitative data. The amount or quantity of a subject or topic under investigation.

Qualitative data. The individual components present in a subject or topic under investigation.

Unit. The physical property and size of a measurement.

Metric system. A decimal system of weights and measurements that is used by most of the world with the exception of the United States.

Systéme Internation (SI). Units of kilograms, meters, cubic meters, seconds and kelvin to represent mass, length, volume, time, and temperature, respectively.

Mass. A measure of how much material or matter is present in an object. This quantity does not change with location.

Kilogram (kg). The standard SI unit for mass. 1kg = 2.205 lb

Gram (g). The standard metric system unit for mass that is one one thousandth of a kilogram. 1 oz = 28.4 g

Weight. A measure of the gravitational pull on an object. This quantity can change with location.

Meter (m). The standard unit for length in both the SI and metric systems. 1 m = 1.1 yd

Cubic meter (m^3). The standard unit for volume in the SI. 1 m^3 = 264 gal

Liter (L). The standard metric unit for volume. 1 L = 1/1000 m^3 = 1.07 qt

Second (s). The standard unit of time in the SI, metric, and English systems. 1 s = 1/60 min

Prefix. One or more letters that go before the name of a standard unit and change the quantity of that unit by a fixed value.

Milliliter (mL). A unit of volume in the metric system. 1 mL = 0.001 L = 1 cm^3 = 1 cc

Cubic centimeter (cm^3 or cc). The commonly used medical term to represent a volume of fluid equal to 1 mL.

Centimeter (cm). A commonly used term in the metric system that represents 0.01 of a meter. 100 cm = 1 m

Systematic error. An inaccurate measurement resulting from a poorly calibrated instrument.

Indeterminate error. Inconsistent data due to random variation in that which is being measured.

Accuracy. The term used to describe how close a measurement is to the true value of the property being measured.

Precision. An indicator of how similar a set of measurements are to one another regardless of how close they are to the true value.

Significant figures. All of the digits shown by a measurement device plus the estimated digit.

Exact numbers. Numbers that have no uncertainty and are assumed to have an unlimited number of significant figures.

Scientific notation. A method to represent a number as units between 1 and 9 multiplied by a power of 10.

Dimensional analysis. A systematic approach for solving both conversions and calculations.

Conversion factors. The ratios or fractions used in dimensional analysis that are derived from definitions or equalities.

Matter. Anything that has mass and occupies space.

Vacuum. The absence of matter.

Gases. Matter that has no definite volume or shape.

Liquids. Matter that has a definite volume but no definite shape.

Solids. Matter that has both a definite shape and a definite volume.

Physical change. A change that alters the state or form of matter but not its content or composition.

Chemical change. A change in the composition of matter that usually involves a change in energy.

Chemical reaction. A chemical change.

Mixture. Matter of variable composition that can be physically separated into individual components.

Substances. Matter that has a constant composition that cannot be physically separated.

Element. A substance that cannot be broken down into a simpler substance.

Chemical symbol. An alphabetical abbreviation used to represent each element on the periodic table.

Compound. A pure substance that contains two or more elements in fixed proportions.

Density. The ratio mass to volume for a given substance is typically expressed in units of grams per milliliter.

Specific gravity. The density of a sample divided by the density of water (1.000 g/mL).

Energy. The ability to do work.

Kinetic energy. The energy associated with motion.

Heat. A form of kinetic energy that involves the motion or vibration of particles in matter.

Potential energy. Stored energy.

Joule (1 J). The standard SI unit for energy. 1 J = 0.2390 cal

Calorie (1 C). The unit used to measure energy in foods in the United States. 1 C= 1 kcal = 1000 cal

Temperature. A measure of the kinetic energy in matter.

Fahrenheit scale. A standard temperature scale used in the United States.

Celsius scale. The standard temperature scale used in the metric system.

Kelvin scale. The standard temperature scale used in the SI.

Absolute zero. The temperature in the kelvin scale where particles in matter are motionless.

Chapter 1 Objective Details:

a. Using Prefixes: Prefixes provide a fixed value change.

Prefix	Abbreviation	Value of Prefix	Prefix in Scientific Notation
Mega	M	1,000,000	1×10^{6}
Kilo	k	1,000	1×10^{3}
Deci	d	0.1	1×10^{-1}
Centi	c	0.01	1×10^{-2}
Milli	m	0.001	1×10^{-3}
Micro	μ	0.000001	1×10^{-6}
Nano	n	0.000000001	1×10^{-9}

Prefixes are typically used to express measurements of volume or length. A common jogging distance would be expressed as 5 km (kilometers) rather than 5,000 m (meters). A person could drive 70 km/hr (kilometers per hour) but you would not typically express this as 70,000 m/hr (meters per hour). The distance between two atoms in a molecule could be 0.5 nm (nanometers) but would not be referred to as 0.0000000005 m (meters). A patient could be given a 10 cc (cubic centimeter) or 10 mL (milliliter) injection, but this would not be referred to as a 0.01 L injection.

b. Scientific Notation: The method by which very large or very small numbers are expressed as a number between 1 and 9 multiplied by a power of 10. When the decimal point is moved to the right, the exponent is given a negative value. When the decimal point is moved to the left, the exponent is given a positive value.

$0.001 = 1 \times 10^{-3} = 1 \times 0.001$
$0.01 = 1 \times 10^{-2} = 1 \times 0.01$
$0.1 = 1 \times 10^{-1} = 1 \times 0.1$
$1 = 1 \times 10^{0} = 1 \times 1$
$10 = 1 \times 10^{1} = 1 \times 10$
$100 = 1 \times 10^{2} = 1 \times 100$
$1000 = 1 \times 10^{3} = 1 \times 1000$

If a number already has an exponential term, the same rules for moving the decimal point apply. When the decimal point is moved to the right, the number of places it is moved is subtracted from the existing exponent. When the decimal point is moved to the left, the number of placed it is moved is added to the exponential term.

$17735 \times 10^{5} = 1.7735 \times 10^{9}$

The decimal point is assumed to be behind the last number prior to the exponential term. In this case, the decimal point was moved four places to the left and the exponent was increased by four units to nine.

$0.000234 \times 10^{9} = 2.34 \times 10^{5}$

For the first number of the expression to be between 1 and 9, the decimal point was moved 4 places to the right and the exponent decreases from 9 to 5.

$0.00739 \times 10^{-3} = 7.39 \times 10^{-6}$

The decimal point was moved 3 places to the right resulting in 3 being subtracted from the existing exponent. Since the exponential was already negative, the subtraction of the additional 3 results in a total exponent of negative 6.

c. Dimensional analysis: A systematic approach to solving both conversions and calculations. There are many instances when the units you are given are not the ones you desire. For those cases, dimensional analysis provides a method for converting between units without changing the integrity of the data.

Example: Convert 75 kilograms to pounds.

Solution:

1.) You must first know the conversion factor or number of pounds per one kilogram. There are 2.2 pounds in one kilogram.
2.) Set up the problem by writing down the value and units of the expression you are given (75 kg). It is then a good idea to write down the units that you are converting to for the answer (lbs). Using a long line may prove helpful to cancel units. The goal is to remove kg from the expression and replace them with lbs. To do this, use the conversion factor and put the units of kg below the line (denominator) and put the corresponding value of 2.2 lbs on the top of the line (numerator).
3.) Multiply 75 kg by 2.2 lbs/1 kg to get the answer to the problem in pounds.
4.) The answer can only be expressed to two significant figures.

$$\frac{75\ \cancel{\text{kg}}}{} \left| \frac{2.2\ \text{lbs}}{1\ \cancel{\text{kg}}} \right. = 170\ \text{lbs}$$

Example: A 50.0 kilometer run is being hosted and you want to know how many miles the run will be.

Solution:

1.) Given 50.0 km

2.) Converting to miles (mi)
3.) Conversion factor: 1 mi = 1.609 km

50.0 km	1 mi	= 31.1 mi
	1.609 km	

4.) Multiply 50 km by 1 mi/1.609 km will cancel kilometers and give the answer in miles.

Dimensional analysis problems will often require several steps to get to the desired unit.

Example: Determine how many milliliters of water are contained in a gallon jug.

Solution.

1.) Given 1 gallon
2.) Convert to milliliters (mL) using the conversion factors: 1 gal = 3.785 L; 1 L = 1000 mL
3.) When you set up the problem, begin with what you are given and converted to a common unit that can then be converted to the desired unit. The only tools you have to work with are conversion factors. For this problem, gallons must be converted to liters before a comparison can be made to milliliters. Here is what it should look like. Notice that the final answer has been presented in two different ways. Both representations are correct, but the latter, which uses scientific notation, is more commonly observed.

1 gal	3.785 L	1000 mL	= 3785 mL = 3.785 x 10^3 mL
	1 gal	1 L	

d. Density: The ratio of a substance's mass to its volume that is typically expressed in units of grams per milliliter.

D = m/v

Density problems typically lend themselves to multi-step dimensional analysis problems. The density of water is 1.00 g/mL.

Example: Express the density of water in terms of pounds per gallon.

Solution: The things to remember for this type of problem are to convert what you are given to common units using the appropriate conversion factors. Convert one unit at a time. Starting with grams, convert to pounds. Once you have one set of terms in the appropriate units, convert milliliters to gallons.

1.) Given D = 1.00 g/mL for water.
2.) Convert to lbs/gal using the conversion factors: 1 lb = 2.2 kg; 1 kg = 1000 g; 1 gal = 3.785 L; 1 L = 1000 mL

1.00 g	1 kg	2.2 lbs	1000 mL	3.785 L	= 8.33 lb/gal
1 mL	1000 g	1 kg	1 L	1 gal	

e. Significant Figures in Calculations.

Until this point, all of the calculations have been performed without the use of significant figures. Significant figures are the number of digits that are known with certainty from a measurement or calculation plus one digit that is uncertain. For the remainder of this study guide, significant figures will be used in each and every calculation. Every measurement made by a scientist in the research or medical field must be carefully recorded using only the information that is known. To evaluate the data, some common rules need to be followed.

Rules for Significant Figures

1. All nonzero digits are significant.
2. Leading zeros are never significant.
3. Confined zeros are always significant.
4. Trailing zeros are significant if a decimal point is present in the number.
5. Trailing zeros are not significant if the number lacks an explicitly shown decimal point.

Rules for Calculations Using Significant Figures

1. For multiplication and division, the measurement containing the fewest significant figures determines the number of significant figures in the answer.
2. For addition and subtraction, the calculated answer should end at the same decimal place as the measurement that has the least precision.

When dealing with significant figures, the question of exact numbers typically creeps into the discussion. An exact number has no uncertainty and is assumed to have an unlimited number of significant figures. Some common examples would be: 1 L = 1000 mL; 1 kg = 1000 g; 1 nm = 1×10^{-9} m; etc.

A simple way to determine the number of significant figures is to write the numbers in scientific notation keeping the above rules in mind.

432 = 3 significant figures: all nonzero digits are significant.

0.0023 = 2 significant figures. When written in scientific notation, 0.0023 = 2.3×10^{-3}. The leading zeros are not significant.

1.00001 = 6 significant figures: confined zeros are always significant.

2.749500 = 7 significant figures: trailing zeros are significant if a decimal point is present in the number.

1,000,000 = 1×10^{6} = 1 significant figure: trailing zeros are not significant if the number lacks an explicitly shown decimal point.

$2.000 \times 0.200 \times 0.20 = 0.08$ = 2 significant figures: In multiplication, the answer is limited to the number of significant figures contained in the value with the lowest number of significant figures. In this case, 0.20 contains 2 significant figures. It is this term that is the limiting factor in the answer. The answer expressed in proper scientific notation and the correct number of significant figures is 8.0×10^{-2}.

$(3.000 \times 6.53) \div 13.567 = 1.44$ = 3 significant figures. In division, the answer is limited to the number of significant figures contained in the value with the lowest number of significant figures. In this case, 6.53 contains 3 significant figures. The answer is then limited to 3 significant figures.

$235.45 + 37 + 36.4 = 309 = 3.09 \times 10^{2}$ = 3 significant figures. In addition, the calculated answer should end at the same decimal place as the measurement that has the least precision. In this case, 37 does not contain any numbers to the right of the decimal point and thus the answer cannot contain any numbers to the right of the decimal.

$3.111 - 2.07 = 1.04$ = 3 significant figures. In subtraction, the calculated answer should end at the same decimal place as the measurement that has the least precision. In this case, 2.07 contains two numbers to the right of the decimal and thus the answer is limited to two numbers to the right of the decimal.

f. Temperature Conversions

$$T_F = 1.8(T_C) + 32 \quad T_C = \frac{(T_F - 32)}{1.8} \quad T_K = T_C + 273$$

T_F = the temperature recorded in degrees Fahrenheit.
T_C = the temperature recorded in degrees Celsius.

T_K = the temperature recorded in Kelvin. *Measurements in Kelvin do not contain the degree designation.

Example: If a patient has a temperature of 40.2°C, what is their temperature on the Fahrenheit and Kelvin scales?

Solution: Fahrenheit scale.

1.) $T_F = 1.8(T_C) + 32$ Replace T_C with the temperature in degrees Celsius.

2.) $T_F = 1.8(40.2°C) + 32$ Multiply 40.2 by 1.8 and then add 32 to the answer.

3.) $T_F = 104.36°F$ Consider the number of significant figures in the measurement and round the answer accordingly. The temperature measurement of 40.2°C contains 3 significant figures and one number to the right of the decimal place. The last operation performed was addition, therefore the answer should be expressed with one value to the right of the decimal place. The correct answer is 104.4°F. Notice that the 3 rounded up to a 4 because the number following it (6) was greater than or equal to 5.

Solution: Kelvin scale.

1.) $T_K = T_C + 273$

2.) $T_K = 40.2°C + 273$

3.) $T_K = 313.2$ K This answer is expressed in the correct number of significant figures because there is only one number to the right of the decimal point in both the measurement and the answer.

Solutions to Odd Numbered Problems:

1. a. The three steps of the scientific method are collection of data, formation of hypotheses, and experimentation to test the hypotheses.
   b. A scientist presented with a problem will first research all known information regarding the problem to be solved. Once the data is collected, a tentative hypothesis or explanation of the problem is formulated. The final step is the experimentation used to test the hypothesis.

3. a. A law is a generalization or summary that is based on consistent experimental results or experiences.
   b. A hypothesis makes predictions that can be tested by experimentation where a law is formulated based on consistent experimental results. A theory is a statement based on one or more accepted hypotheses.

5. A hypothesis can never be proved correct; experimentation can only show it to be acceptable or invalid.

7. a. The components of a measurement are a number and a unit.
   b. A unit specifies the physical property and size of the measurement, and the number indicates how many units are present.

9. a. meter
   b. kilogram
   c. cubic meter
   d. second

11. a. 14 g is the abbreviation for 14 grams which is a measurement of mass.
    b. 16.3 s is the abbreviation for 16.3 seconds which is a measurement of time.
    c. 47.3 L is the abbreviation for 47.3 liters which is a measurement of volume.
    d. 385 m is the abbreviation for 385 meters which is a measurement of length.

13. The weight of an object is dependent on the gravitational pull exerted on that object where the mass of the object is not affected by gravity.

15. a. centi- 0.01
    b. milli- 0.001
    c. kilo- 1000

17. a. centi- c
    b. milli- m

c. kilo- k

19. a. 64.1 mL = 64.1 milliliters = 0.0641 liters = volume
    b. 4.3 ns = 4.3 nanoseconds = 0.0000000043 seconds = time
    c. 23 Mg = 23 megagrams = 23000000 grams = mass
    d. 34.32 cm = 34.32 centimeters = 0.3432 meters = length

21. a. 1 cc = 1 mL
    b. 1 L = 1000 cc

23. a. 1000 m = 1 km
    b. 1 s = 1000000 μs
    c. 1000000 g = 1 Mg

25. The measurement from a barometer that is marked to the nearest millimeter should be recorded to one decimal position (tenths place). The unit markings are to the whole number or millimeter place and then you need to approximate the position between the whole number places for the measurement.

27. a. Precision is an indicator of how similar a set of measurements are to one another regardless of how close they are to the true value.
    b. A fine watch will have the ability to make measurements that are very close together making it more precise than a sundial.

29. A significant figure represents all of the digits shown by a measurement device plus an estimated digit.

31. a. 17 km has 2 significant figures because all nonzero numbers are significant.
    b. 12.417 g has 5 significant figures because all nonzero numbers are significant.
    c. 68.4 s has 3 significant figures because all nonzero numbers are significant.

33. a. 83,000 mi has 2 significant figures because trailing zeros are not significant if the number lacks an explicitly shown decimal point.
    b. 138.006 μm has 6 significant figures because confined zeros are always significant.
    c. 0.000005 L has 1 significant figure because leading zeros are never significant.
    d. 340.2100 cm has 7 significant figures because trailing zeros are significant if a decimal point is present in the number.
    e. 1,000,000 km has 1 significant figure because trailing zeros are not significant if the number lacks an explicitly shown decimal point.

35. For multiplication and division, the measurement containing the fewest significant figures determines the number of significant figures in the answer. For addition and subtraction, the calculated answer should end at the same decimal place as the measurement that has the least precision.

37. a. 214 When the numbers are correctly typed into your calculator and the multiplication function performed, the answer the calculator gives you is 213.864. The number 93,800 contains only 3 significant figures because trailing zeros are not significant if the number lacks an explicitly shown decimal point. The answer to the problem can therefore only contain three significant figures. The fourth number in the answer is an 8 meaning that the 3 must be rounded to a 4 making the answer 214.
    b. 70000 The answer your calculator will give you for this multiplication problem is 68,568. The correct number of significant figures is limited to 1 in the answer because of the 6 in the problem. The 68,568 must be rounded to give 1 significant figure resulting in the first number being rounded up because the second number is greater than 5.
    c. 270 The answer your calculator will give you is 274.56756. The correct number of significant figures for division is limited by the measurement containing the fewest significant figures. The 11 has only two significant figures so the answer is limited to two significant figures. The number 274.56756 rounded to two significant figures is 270 or 2.7×10^2.

d. 476 The answer you calculator will give you is 476.1046. The number 5280 contains only three significant figures, so the answer to the problem is limited to three significant figures.

39. a. 43008 The answer to this problem cannot contain any numbers to the right of the decimal point because addition problems are limited to the measurement that has the least precision.
    b. 1.845 The calculated answer to this problem is 1.84477. The answer to the correct number of significant figures is limited to three digits to the right of the decimal point. The third digit to the right of the decimal point is a four that is rounded to a five because it is followed by a seven that is greater than five.
    c. –4.30 The answer is limited to two digits to the right of the decimal place. The calculated answer is –4.299. The correct rounding to two digits to the right of the decimal place yields the correct answer to the correct number of significant figures.
    d. 11.346 The answer is limited to three digits to the right of the decimal place.

41. a. $32{,}110 = 3.2110 \times 10^4$
    b. $0.0000000005598 = 5.598 \times 10^{-10}$
    c. $1008 = 1.008 \times 10^3$
    d. $0.000200 = 2.00 \times 10^{-4}$

43. a. $1.77 \times 10^{11} = 177000000000$
    b. $2.92 \times 10^4 = 29200$
    c. $6.58 \times 10^{-3} = 0.00658$
    d. $9.477 \times 10^{-7} = 0.0000009477$

45. a. 1×10^9 ns/1 s or 1 s/1×10^9 ns
    b. 1000 g/1 kg or 1 kg/1000 g
    c. 1 L/1.057 qt or 1.057 qt/1 L

47. a. 1 in = 2.54 cm
    b. 2.205 lb = 1 kg
    c. 1000 m = 1 km

49. Dimensional analysis involves canceling units to convert to desired units.

a. $\frac{(33.5\ \cancel{in})\ (1\ ft)}{(12\ \cancel{in})} = 2.79\ ft$ b. $\frac{(7.3\ \cancel{pt})\ (1\ qt)}{(2\ \cancel{pt})} = 3.7\ qt$ c. $\frac{(3.770\ \cancel{lb})\ (16\ oz)}{(1\ \cancel{lb})} = 60.32\ oz$

51.

a. $\frac{807\ \cancel{cm} \mid 1\ m}{\mid 100\ \cancel{cm}} = 8.07\ m$ b. $\frac{43\ \cancel{g} \mid 1\ kg}{\mid 1000\ \cancel{g}} = 0.043\ kg$ c. $\frac{2.31\ \cancel{L} \mid 1000\ mL}{\mid 1\ \cancel{L}} = 2310\ mL$

53.

a. $\frac{2.7\ \cancel{kg} \mid 2.2\ lb}{\mid 1\ \cancel{kg}} = 5.9\ lb$ b. $\frac{100\ \cancel{yd} \mid 1\ m}{\mid 1.094\ \cancel{yd}} = 90\ m$ c. $\frac{37.4\ \cancel{L} \mid 1.057\ qt}{\mid 1\ \cancel{L}} = 39.5\ qt$

55.

a.

126 μL	1×10^{-6} L	1000 mL	= 0.126 mL
	1 μL	1 L	

b.

3.5 cups	1 pt	1 qt	= 0.88 qt
	2cups	2 pt	

c.

163 μm	1×10^{-6} m	100 cm	2.54 in	= 4.14×10^{-2} in
	1 μm	1 m	1 cm	

57. 1 kg = 2.2 lb

2.25 kg	2.2 lb	= 4.95 lb
	1 kg	

59. 1 oz = 28.4 g

3.0 oz	28.4 g	= 85 g
	1 oz	

61.

a.

30 days	3000 mg	1 tablet	= 180 tablets
	1 day	500 mg	per month

b.

180 tablets	17 cents	1 dollar	= 30.60 dollars
1 month	1 tablet	100 cents	per month

63. a. A vacuum is the absence of matter.
    b. A perfect vacuum is the absence of any and all matter. On Earth, everything around us consists of matter. There are no perfect vacuums in nature.
    c. Outer space is not a perfect vacuum because tiny amounts of matter are present in outer space.

65. Liquid

67. A solid has a definite shape and occupies a definite volume.

69. a. Physical
    b. The physical separation of coffee from grounds can be accomplished by filtration or decanting.
    c. Yes!

71. a. A substance is matter that has a constant composition that cannot be physically separated.
    b. Three examples of substances are table salt, gold and copper.

73. a. A compound is a pure substance that contains two or more elements in fixed proportions.
    b. Three examples of compounds are table salt, water, and hydrogen peroxide.

75. a. O
    b. Au
    c. S
    d. U
    e. Na

77. a. Potassium
    b. Copper
    c. Calcium
    d. Nitrogen
    e. Chlorine

79. Density is the ratio of a substances mass to its volume that is typically expressed in units of grams per milliliter. The specific gravity of a substance is a unit-less expression of the density of that substance divided by the density of water.

81. a. Density is equal to the mass of the substance divided by the volume of that substance. Dividing 723.4 g by 68.6 mL and keeping in mind significant figures, the answer is 10.5 grams per milliliter.
    b. A solid will have a definite volume and a definite mass. The volume of the cube can be determined by multiplying length times height times width. If the measurements of the length, height, and width are made in centimeters, then the volume will be determined in units of cubic centimeters that are equivalent to milliliters. Dividing the mass of the cube by the volume of the cube will yield the density.

83. Dimensional analysis is the best way to solve this problem.

150.0 mL	1.12 g	
	1 mL	= 168 g

85. Energy is the ability to do work.

87. a. Heat is a form of kinetic energy that involves the motion or vibration of particles in matter.
    b. Heat will cause particles of matter to move or vibrate faster.

89. a. Unit conversions: 1J = 0.2390 cal; 1 C = 1 kcal = 1000 cal
    1.00×10^2 J = 0.100 kilojoules = 23.90 calories = 0.02390 kilocalories = 0.02390 Calories
    b. Energy in foods in the United States is expressed in units of Calories.

91. a. Kelvin system = Kelvin (K); Celsius = degree Celsius (°C); Fahrenheit = degree Fahrenheit (°F)
    b. degree Fahrenheit
    c. Kelvin and Celsius degrees are the same size.

93. a. 611.1°C
    b. 39.4°C
    c. -194°C

95. a. 223°F
    b. 40°F
    c. –459.67°F

97. a. 329K
    b. 536K
    c. 190K

99. a. 100°C
    b. 556°C
    c. -157°C

101.a. The large difference in price can be attributed to the additional three significant figures that can be measured with the balance that goes to ten-thousandth of a gram.
    b. You would need to know how large the quantities are that are currently being measured by the lab balances.

103.

4 tsp	1 tbl	0.5 oz	29.6 mL	21 days	
1 day	3 tsp	1 tlb	1 oz	perscription	= 414.4 mL per perscription

Sample Test:

1. Which of the following is a compound?
   a. acetaminophen ($C_8H_9NO_2$) b. silver c. wine d. argon

2. Which of the following is a mixture?
   a. acetaminophen ($C_8H_9NO_2$) b. silver c. wine d. argon

3. The fact that gold does not rust is a
   a. physical property b. personal property c. real property d. chemical property

4. Which of the following represents a physical change in matter?
   a. a substance solidifies when cooled
   b. a substance gives off a gas when heated
   c. a substance burns when heated
   d. a substance remains unchanged as it is heated

5. Which of the following is a chemical property of matter?
   a. color b. density c. freezing point d. flammability

6. The limit of chemical subdivision of a pure substance consists of
   a. atoms b. molecules c. protons d. compounds

7. Two pure substances A and B react to form a new pure substance C. From this we may conclude that
   a. A and B are both elements
   b. C is a compound, A and B may or may not be elements
   c. C is an element, A and B are compounds
   d. A, B, and C are all compounds

8. One millionth (10^{-6}) of a basic unit in the metric system is indicated by the prefix _______________?

9. The base unit of mass in the metric system is the _____________ and the base unit of volume in the metric system is the _____________.

10. Give the name of the element that corresponds to the following symbols.
    Ag, Be, C, Ca, Fe, H, K, Na

11. The three steps to the scientific method are:

12. Rank the following units in order of increasing size.
    a. 1 km
    b. 1 µm
    c. 1 cm
    d. 1 mm

13. Circle the letters corresponding to all of the following terms that apply to the substance containing only H_2O molecules.
    a. element b. compound c. pure substance d. mixture

14. Which of the following is a physical property of matter?
    a. it does not burn
    b. it fizzes with acid
    c. the freezing point is -10°F
    d. the surface turns black in air

15. As two clear liquid solutions are added together, a red solid forms. This change is most likely
    a. physical b. chemical c. Neither chemical not physical d. both chemical and physical

16. The limit of physical subdivision of pure H_2O is
    a. the atom b. the molecule c. the element d. a proton

17. One hundredth (10^{-2}) of a basic unit in the metric system is indicated by the prefix _____________?

18. Multiply (2.3×10^{-6})(5.0×10^{2}) and express the answer in standard scientific notation with the correct number of significant figures.

19. Add 6.0 + 2.47 + 0.004 and express the answer using correct significant figures.

20. If 1 inch = 2.54 cm, what is the length, in meters, of an object that is 6.5 inches? Give the answer in standard scientific notation and the correct number of significant figures.

21. If a person can comfortably carry 20 kg (44 lbs) and gold has a density of 19.6 g/cm^3, what is the volume of the largest piece of solid gold a person can carry? Give the answer in standard scientific notation and the correct number of significant figures. (don't forget the units)

22. Convert a temperature of 24°C to a Fahrenheit value and a Kelvin value.

23. A hospital patient has a fever of 103.5°F and weights 175 lb. He is to receive medication for his fever, the dosage of which is 40.0 mg per kg of body mass. The drug is dissolved in water at a concentration of 20.0 mg per mL of water.
    a. What is the patient's body temperature in degrees Celsius?
    b. What is the patient's mass in kilograms? (1 lb = 454 g)
    c. What mass, in milligrams, of pure drug should the patient receive?
    d. How many milliliters of the drug solution should the patient be given?

24. Divide (4.7×10^{3})/(6.1×10^{-4}) and express the answer in standard scientific notation and the correct number of significant figures.

25. Subtract 0.345 – 0.0221 and express the answer using correct significant figures.

26. A vitamin preparation weighs 255 milligrams. Express this weight in kilograms useing standard scientific notation.

27. Nickel metal has a density of 8.90 g/cm^3. How much does 15 cm^3 of nickel metal weigh in grams? Give the answer in standard scientific notation and the correct number of significant figures. (don't forget the units)

28. You weigh yourself before going to the doctors office and your $12 Walmart special scale shows that you are 135 lbs. a.) Have you determined your weight or your mass? Explain! b.) If the doctors $1200 scale indicates that you are 153.45 lbs, what can you determine about your scale in terms of accuracy and precision?

29. Which is the higher temperature, -15°C or 4°F?

30. What are the three states of matter and how are they different?

Answers to Sample Test:

1. a
2. c
3. d
4. a
5. d
6. a
7. b
8. micro
9. gram; liter
10. silver; beryllium; carbon; calcium; iron; hydrogen; potassium; sodium
11. data collection; hypothesis formation; experimental testing of hypothesis

12. b, d, c, a
13. b, c
14. c
15. b
16. b
17. centi
18. 1.2×10^{-3}
19. 8.5
20. 1.7×10^{-1} m
21. 1×10^{3} cm^{3}
22. 75°F; 3.0×10^{2}K
23. a. 39.7°C
    b. 79.5 kg
    c. 3.18×10^{3} mg
    d. 159 mL
24. 7.7×10^{6}
25. 0.323
26. 2.55×10^{-4} kg
27. 1.3×10^{2} g
28. a. Weight: You have determined a value that is dependent on the gravitational pull of the earth.
    b. Your home scale is neither accurate nor is it as precise as the one in the doctors office.
29. -15°C
30. Matter exists as either a solid, liquid, or gas. Gases have no definite volume or shape. Liquids have no definite shape, but do have a definite volume. Solids have both a definite shape and a definite volume.

Chapter 2: The Atom

Chapter Objectives:

1. To understand that all matter is made up of individual atoms of one or more elements.
2. To gain a broad understanding of subatomic particles and how they can be used to identify elements, isotopes, and/or ions.
3. To distinguish between valence electrons and interior electron configurations.
4. To understand the groups and periods of the periodic table.
5. To learn the filling order of subshells that make up the electron configuration about the nucleus of an atom.
6. To understand the relationship between nuclear radiation and the different types of radioactive decay.
7. To establish how radiation is detected and expressed.
8. To learn the concept of nuclear reactions and their uses.

Key Terms: The key terms are listed in the order in which they are encountered in the Chapter.

Chemistry. The study of matter and of the changes that occur in matter.

Atom. The smallest particle of an element that possesses the properties of that element.

Periodic Table. A table listing all of the elements arranged in order of increasing atomic number.

Subatomic particles. The electrons, protons, and neutrons that give an atom its identity.

Electron. A subatomic particle of negligible mass and an electrical charge of negative one.

Proton. A subatomic particle that has a mass of one atomic mass unit and a charge of plus one.

Neutron. A subatomic particle that has a mass of one atomic mass unit and is neutral in charge.

Atomic mass unit (amu). One-twelfth the mass of a carbon-12 atom or the mass of a proton and/or neutron.

Nucleus. The positively charged core of an atom that contains protons and neutrons.

Neucleon. The protons and neutrons at the core or nucleus of an atom.

Ion. An atom that does not contain the same number of protons as electrons.

Atomic Number. The number of protons found in any atom of an element.

Isotopes. Atoms of an element that have different numbers of neutrons.

Mass number. The sum of the neutrons and protons found in the nucleus of an atom.

Atomic mass. The average mass of the naturally occurring isotopes of an element expressed in atomic mass units.

Shells. Energy levels or layers of electrons that contain increasing number of electrons the further from the nucleus they are found.

Valence shell. The outer most shell of electrons in an atom.

Valence electrons. The electrons that occupy the outer most shell around the nucleus of an atom.

Core electrons. All of the electrons in an atom that occupy shells other than the valence shell.

Electromagnetic radiation. Energy that is emitted and propagated in the form of rays or waves.

Wavelength. The length of the wave.

Frequency. The number of waves of light that pass a point in a second.

Periodic law. The properties of the elements vary in a systematic way according to outer electronic arrangement.

Groups. Columns on the periodic table that are arranged based on their number of valence electrons.

Alkali metals. Elements in Group IA that react with water to make them basic.

Alkaline earth metals. Elements in Group IIA that are common in rocks and minerals.

Halogens. Elements in Group 7A that typically exist as salts when in the compound form.

Noble gases. Elements in Group 8A that are less reactive than other elements.

Metals. Elements that are shiny, malleable, ductile, and are good conductors of heat and electricity.

Nonmetals. Elements that are dull, brittle, and electrical insulators.

Metalloids. Semimetals known as semiconductors because of their ability to conduct electricity somewhat.

Period. Rows of elements in the periodic table that have their valence electrons in the same shell.

Subshell. A subdivision of electron shells that are labeled s, p, d, and f in order of increasing energy.

Orbital. The region in space within a subshell where a maximum of two electrons can be found.

Nuclear radiation. Radiation that originates in the nuclei of atoms.

Radioactive nuclides. An atom that has an unstable nucleus that changes into an atom with a different nucleus with time.

Radioactive decay. The process by which an unstable or radioactive nucleus emits a particle that changes either the mass number or atomic number of the nucleus.

Half-life. The time required for one half of a sample of a radioactive nuclide to decay.

Nuclear radiation. The energy and mass emitted by radioactive nuclides as they decay into more stable nuclides.

Alpha radiation. Radiation that consists of alpha (α) particles.

Alpha particle. A heavy particle with two protons and 2 neutrons represented by the helium-4 symbol.

Beta radiation. Radiation that consists of beta (β) particles.

Beta particle. An electron with no appreciable mass and an electronic charge of negative one.

Gamma radiation. Radiation consisting of gamma (γ) particles.

Gamma particle. A high-energy particle with no appreciable mass.

Nuclear equation. An equation that represents the decay of a radioactive atom into a more stable nuclide.

Daughter nuclide. The nuclide formed when an unstable nuclide undergoes decay.

Ionizing radiation. Radiation that strikes an atom with enough energy to cause the atom to release an electron.

Geiger-Müller tube. A gas-filled tube that detects ionizing radiation to be recorded by a Geiger counter.

Scintillation detector. A device that measures the amount of light produced from the interaction of light and matter.

Curie (Ci). The unit of radiation equal to 3.7×10^{10} disintegrations per second (dps).

Disintegrations per second (dps). The number of nuclear decays that occur in a period of one second.

Radiation absorbed dose (rad). The units used to express the amount of energy associated with radiation. One rad is equal to 1×10^{-2} J of energy being absorbed by 1 kg of exposed tissue.

Radiation equivalent for man (rem). A unit of radiation used to represent the ability of the radiation to damage tissues. One rem is one rad multiplied by the relative biological effectiveness factor (RBE).

Relative biological effectiveness factor (RBE). This factor is assigned a value of 1 for beta and gamma radiation and 20 for alpha particles produced in the body.

LD_{50}. A lethal dose for 50% of the population.

Background radiation. Radiation from natural sources that is always present in our environment.

Fission. The splitting of a large nucleus into smaller nuclei with the release of enormous amounts of energy.

Subcritical fission. A fission reaction in which, on average, fewer than one of the product neutrons from a fission event produces another fission event.

Critical chain reaction. On average, one neutron from each fission event produces another fission event.

Supercritical chain reaction. On average, more than one neutron from each fission event causes another fission event causing the number of fission events to increase rapidly.

Fusion. The combining or fusing of smaller nuclei into larger ones to produce enormous amounts of energy.

Nuclear transformation. The conversion of one nucleus into another by high-energy bombardment with a particle.

Chapter 2 Objective Details:

a. The atom. Atoms are the smallest particles that possess the properties of an element. Conversely, an element can be broken down no further than an atom. Atoms are made up of subatomic particles called protons, neutrons, and electrons. For all practical purposes, protons have a charge of plus one and a mass of one, neutrons have no charge and a mass of one, and electrons have no mass and a charge of negative one. It makes sense then that protons will attract electrons since they possess opposite charges. In the structure of an atom, all of the mass is in the nucleus or core. Thus the nucleus is comprised of the protons and neutrons. The electrons occupy shells of increasing energy around the nucleus of the atom. These shells are roughly comparable to the periods on the periodic table. The first shell contains one orbital that is called the 1s subshell. An orbital can contain a maximum of two electrons. The electrons occupying the outermost shell are known as the valence electrons. The valence electrons are roughly the same number as the group number of the element of interest on the periodic table. If an element is in group 5A, it will have 5 valence electrons. To put it all together, there is a maximum of seven shells possible. Each new shell accommodates more electrons than the last one. Inside of each shell are subshells that are given the designation s, p, d, and f. The s subshell contains one orbital and a maximum of 2 electrons. The p subshell contains three orbitals and a maximum of 6 electrons. The d orbitals contain five orbitals and a maximum of 10 electrons. The f orbitals contain seven orbitals and can accommodate a maximum of 14 electrons. This information is all contained in the periodic table. If you look closely at the periodic table, groups 1A and 2A are consistent with the s subshell, groups 3A through 8A are consistent with the p subshells, the elements in the center of the table are consistent with the d subshells, and the elements at the bottom are consistent with the f subshells. Consistent means that there are as many elements in each of these sections of the periodic table as there are electrons in each subshell. This is because the elements on the periodic table are arranged according to the number of protons in their nucleus. Since protons attract electrons, the periodic table is also arranged in order of increasing numbers of electrons. Figure 1 is a diagrammatic representation of the regions of the periodic table.

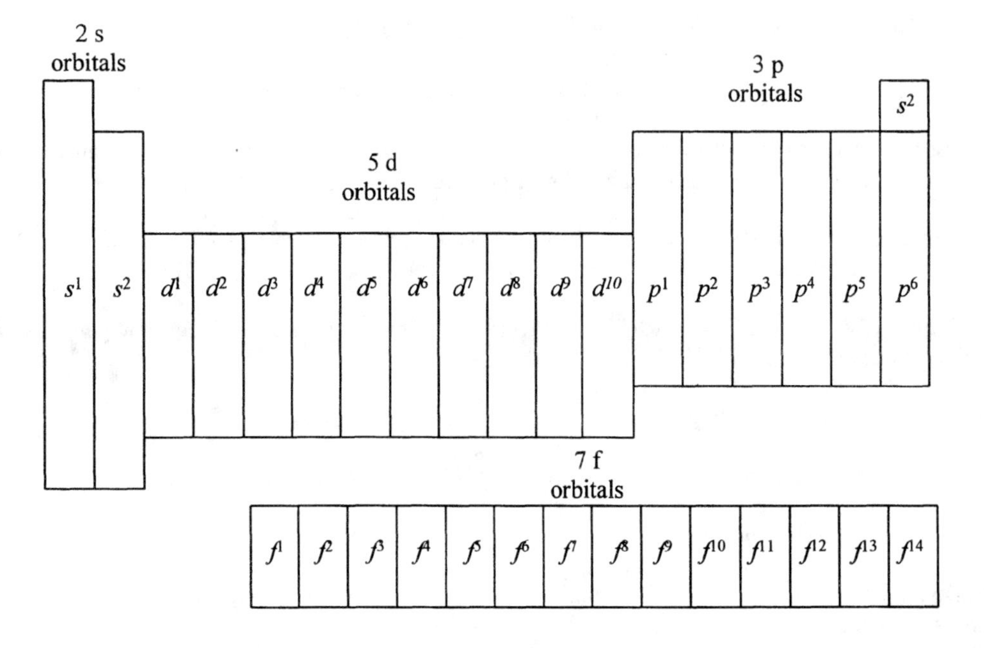

Figure 1
This is a representation of the periodic table showing the subshells s, p, d, and f. The number of electrons in each subshell can be counted, as there are 2 columns in the s region to represent the two electrons of the s subshell. The p subshell can accommodate up to six electrons; thus there are 6 columns in the p region of the periodic table. The same theory holds true for the d subshell, where there are 10 columns and for the f subshell where there are 14 columns.

Figuring out how to fill the shells and subshells for atoms is not trivial. I alluded to the fact that the shell is roughly equivalent to the period that the element occupies in the periodic table. If we consider that each period increases the shell number, then it is consistent to conclude that each period will have an increasing number of subshells as well. The first shell can accommodate only two electrons or more accurately, only two electrons will fill the 1s subshell. If an atom has two electrons, we can determine many things about that atom. The electron configuration will be $1s^2$. The one represents the first shell, the s represents the s subshell, and the 2 are the number of electrons that are in that subshell. The element with two electrons will have two protons as well. Since the periodic table is arranged according to the number of protons an element contains, the atomic number for this element is 2. The element with an atomic number of two is helium. Figure 2 gives a quick and easy way to figure out the order of the shells and subshells.

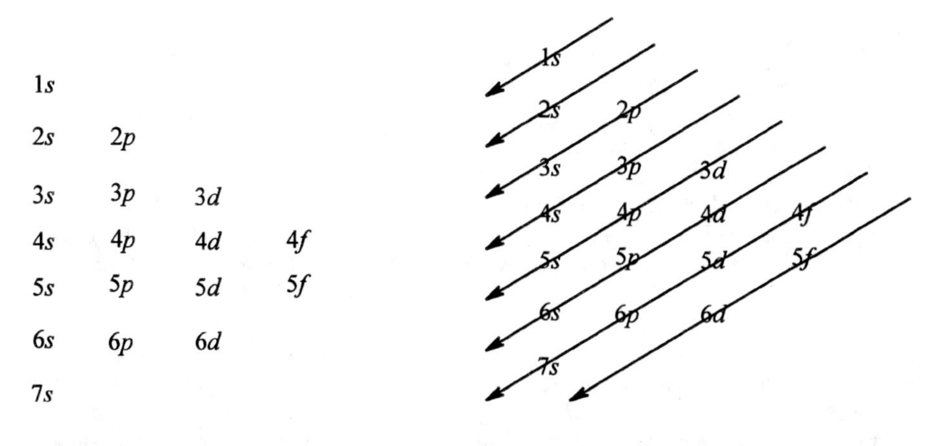

Figure 2
This is a diagrammatic representation to show the order in which subshells should be filled with electrons. The first discrepancy to note is that once the 3p subshell is filled, the 4s subshell is filled prior to the 3 d subshell. This is because the energy of the 4s subshell is lower than that of the 3d subshells.

Let's use bromine as an example for determining electron configurations. First we need to know that the elemental symbol for bromine is Br. If we know the symbol, we can look on the periodic table to find the atomic number.

atomic mass ⟶ $^{80}_{35}$Br ⟵ elemental symbol
atomic number ⟶ 35

Recalling that elements are listed in order of increasing numbers of protons, which is the equivalent to an increasing atomic number, tells us the number of each subatomic particle in an atom of the element. The atomic number for bromine is 35. This means that there are 35 protons in the nucleus of the atom. For a neutral atom of bromine, the number of protons will equal the number of electrons. Thus there are 35 electrons in various shells surrounding the nucleus. The atomic mass tells us the number of protons and neutrons in the nucleus of the atom. Subtracting the number of protons from the atomic mass tells us that there are 45 neutrons in an atom of bromine. Using the chart in Figure 2, we are now ready to determine the electron configuration for bromine. The first shell contains a s subshell that can accommodate 2 electrons ($1s^2$). Once we fill this shell, it is time to move to the second shell. The second shell consists of both s and p subshells. An s subshell contains one orbital and a p subshell contains three orbitals. Recalling that each orbital can contain a maximum of two electrons allows us to put eight electrons into the second shell. This is an additive process meaning that the first two shells will hold a total of ten electrons ($1s^22s^22p^6$). It is now time to move to the third shell. The 3s and 3p subshells can be filled with a total of eight electrons ($1s^22s^22p^63s^23p^6$). If we then follow the periodic table from left to right, we encounter the 4s subshell before the 3d subshells. This is the discrepancy that is being pointed out. The 4s subshell must be filled before the 3d subshells. Two electrons will fill the 4s subshell and ten electrons will be required to fill the 3d subshell ($1s^22s^22p^63s^23p^64s^23d^{10}$). The electron configuration we now have shows a total of 30 electrons. Element 30 is zinc. We still need to include 5 more electrons for the electron configuration of bromine. These electrons will next occupy the 4p subshells. The final electron configuration for bromine is $1s^22s^22p^63s^23p^64s^23d^{10}4p^5$. The highest energy electrons for any atom are the external electrons known as the valence electrons. From the electron configuration the valence electrons are always the electrons from the highest number shell. In the case of bromine, the number of valence electrons is seven ($4s^24p^5$). For the elements in group 1A through 8A, the number of valence electrons is equal to the group number.

Another example would be titanium-48.

1. Look for Ti on the periodic table.

   atomic mass ⟶ $^{48}_{22}$Ti ⟵ elemental symbol
   atomic number ⟶ 22

2. Determine the number of each subatomic particle.
   atomic number = number of protons = 22 protons
   22 protons – 48 protons and neutrons (atomic mass) = 26 neutrons
   22 protons = 22 electrons

3. Determine the electron configuration.
   Read the periodic table like a book. Start from the top left and read to the right. Fill the first shell that contains the s subshell before going to the second shell that contains both s and p subshells. Titanium is in the fourth period, so there will be 4 shells. Always start with the first shell and continually add electrons until each orbital contains two electrons.
   Electron configuration for Ti = $1s^2 2s^2 2p^6 3s^2 3p^6 4s^2 3d^2$
4. Determine the number of valence electrons for titanium.
   Remember the number of valence electrons comes from the electrons in the highest shell. In this case, the fourth shell contains 2 electrons. Do not be fooled by the 3d subshell not being filled. Valence electrons are the electrons in the outermost shell!

b. Nuclear radiation. Nuclear radiation deals with the emission of high-speed particles from the nucleus of an atom. The three types of radiation are alpha, beta, and gamma. Alpha radiation consists of alpha particles that are the equivalent of a helium-4 nucleus. When a radioactive element emits alpha particles, the mass of the original element decreases by 4 atomic mass units and the atomic number or number of protons decreases by 2 units. Beta radiation consists of beta particles that are equivalent to electrons. When a radioactive element emits beta particles, the atomic number of the daughter nucleus increases by one. Beta emission is equivalent to converting a neutron in the nucleus of a radioactive atom to a proton. Thus, the net effect of beta decay is to increase the atomic number by one. Gamma radiation occurs when an atom has excess energy that it must release. There is no appreciable mass or charge changes associated with gamma decay, but a great deal of energy is released.

Atoms of different radioactive elements emit radiation at different rates. A common problem is to calculate the amount of a radioactive isotope that will remain after a certain number of half-lives have passed. For each half-life that passes, divide the original amount of radioactive material by two. The amount of radiation found in a given substance can determined by measuring the number of disintegrations per minute (dpm). Radiation can be detected by photographic films, a Geiger counter, or a scintillation detector. The units for radiation include the curie, radiation absorbed dose (rad), and the radiation equivalent for man (rem).

c. Nuclear reactions. There are three types of nuclear reactions that are often encountered. The most common type of nuclear reaction is fission. Fission deals with the breaking apart of the nuclei of atoms to release large quantities of energy. This is the common process used in nuclear power plants. Fusion combines nuclei to release large quantities of energy. Fusion requires very high temperatures to combine the nuclei making it a less common source of energy. The process of nuclear transformation makes many radioactive elements that do not exist in nature. In this process, nuclei are bombarded with a particle, usually a neutron, and a new nucleus is formed.

<u>**Solutions to Odd Numbered Problems:**</u>

1. An atom.

3. In Dalton's original hypothesis, he proposed that atoms of the same element were the same in all properties including mass. In modern atomic theory, it is understood that different isotopes exist for a single element that is different in mass.

5. a. B
   b. Na
   c. Al
   d. Pt

7. a. Sulfur
   b. Tin
   c. Bromine
   d. Chromium

9. protons, neutrons, and electrons

11. a. proton

b. electron
c. neutron

13. nucleons

15. An atomic mass unit is one-twelfth the mass of a carbon-12 atom. One amu is equal to 1.67×10^{-24} g.

17. The atomic number tells us the number of protons found in the nucleus of each atom of an element.

19. a. 3
b. 15
c. 35
d. 7
e. 12
f. 16

21. a. hydrogen
b. oxygen
c. silver
d. mercury

23. a. Potassium contains 19 protons and 19 electrons.
b. Gold contains 79 protons and 79 electrons.
c. Boron contains 5 protons and 5 electrons.

25. A neutral atom of zinc will have 30 electrons.

27. The mass number of an atom corresponds to the number of neutrons and protons found in the nucleus of an atom.

29. The atomic number is 11 and the mass number is 23.

31. a. 35 protons and 46 neutrons
b. 19 protons and 21 neutrons
c. 8 protons and 10 neutrons
d. 3 protons and 4 neutrons

33.

Element Name	Element Symbol	Atomic Number	Mass Number	Number of Protons and Electrons	Number of Neutrons
Sulfur	S	16	32	16	16
Uranium	U	92	238	92	146
Chlorine	Cl	17	37	17	20
Potassium	K	19	41	19	22

35. The atomic mass of bromine is 79.9 amu because there is a greater abundance of the bromine-79 isotope than there is of the bromine-81 isotope.

37. a. 30.97
b. 22.99
c. 9.01
d. 126.90
e. 39.95

39. a. Tungsten (W)

b. Boron (B)
c. Nickel (Ni)

41. The first shell (1s shell) in an atom is the closest to the nucleus making it the most stable and lowest in energy.

43. a. 8 electrons
b. 32 electrons
c. The maximum number of electrons in a shell can by calculated with the formula $2n^2$, where n is the number of the shell. There are a maximum of 50 electrons in the fifth shell using this formula.

45. a. Aluminum has 2 electrons in the first shell, 8 electrons in the second shell, and 3 electrons in the third shell.
b. Nitrogen has 2 electrons in the first shell and 5 electrons in the second shell.
c. Calcium has 2 electrons in the first shell, 8 electrons in the second shell, 8 electrons in the third shell, and 2 electrons in the fourth shell.

47. a. Hydrogen
b. Magnesium
c. Boron

49. Valence electrons are located in the highest energy outermost shell surrounding the nucleus of an atom.

51. a. An atom of lithium has 1 valence electron and 2 core electrons.
b. An atom of nitrogen has 5 valence electrons and 2 core electrons.
c. An atom of calcium has 2 valence electrons and 18 core electrons.

53. If an atom gains specific amounts of energy, it will jump to higher energy (outer) shells, then it will emit light as it returns to lower energy inner shells.

55. Each of the lines emitted by hydrogen represents the movement of an electron from a higher energy shell to a lower energy shell.

57. The columns of the periodic table are referred to as groups or families.

59. All of the elements of a period have their valence electrons in the same shell.

61. a. Calcium is found in the fourth period.
b. Phosphorous is found in the third period.
c. Carbon is found in the second period.

63. a. Na, Mg, Al, Si, P, S, Cl, Ar
b. H, Li, Na, K, Rb, Cs, Fr
c. C, Si, Ge, Sn, Pb
d. Kr

65. a. 4
b. 3
c. 7

67. The halogens are elements in group 7A that typically exist as salts when in the inorganic compound form.

69. Metals are located to the left on the periodic table.

71. Nonmetals are located to the right on the periodic table.

73. Metalloids are located on the border between metals and nonmetals.

75. The s, p, d, and f subshells can hold a maximum of 2, 6, 10, and 14 electrons, respectively.

77. a. $1s^2$
    b. $1s^22s^22p^1$
    c. $1s^22s^22p^63s^23p^6$

79. An orbital is the region in space within a subshell where a maximum of two electrons can be found.

81. a. 6
    b. 2
    c. 14
    d. 10

83. Nuclear radiation is the energy and mass emitted by radioactive nuclides as they decay into more stable nuclides.

85. An atom that has an unstable nucleus is referred to as a radioactive nuclide.

87. Two half-lives must pass for 12 mg of nickel-66 to decay to 3.0 mg.

89. Two half-lives must pass which is the equivalent of 112 hours.

91. Every 12.5 year's would decrease the amount of tritium by one half. In 50 years there are the equivalent of 4 half-lives that will pass. The 2.50 g sample would be broken down to 1.25 g after 12.5 years, 0.625 g after 25 years, 0.313 g after 37.5 years and after 50 years, there would be 0.156 g remaining.

93. To study the sample of cotton cloth, a measure of disintegrations per minute (dpm) must be performed. It is known that in living plants like cotton, there is enough carbon-14 taken in to release 14 dpm. For each half-life of 5730 years there will be half of the dpm given off. If the cotton cloth were truly 5000 years old, then less than one half life would have passed and an accurate approximation of age would not be possible.

95. Beta radiation consists of the emission of beta particles (electrons). Considering the nucleus of an atom only contains protons and neutrons, beta emission is the equivalent of converting a neutron to a proton. The effect of converting a neutron to a proton is to change the atomic number of the element, increasing it by one.

97.

a. $^{222}_{86}\text{Rn} \xrightarrow{a} {}^{4}_{2}\text{He} + {}^{218}_{84}\text{Po}$

b. $^{210}_{82}\text{Pb} \xrightarrow{b} {}^{0}_{-1}\text{e} + {}^{210}_{83}\text{Bi}$

c. $^{45}_{19}\text{K} \xrightarrow{b} {}^{0}_{-1}\text{e} + {}^{45}_{20}\text{Ca}$

d. $^{238}_{92}\text{U} \xrightarrow{a} {}^{4}_{2}\text{He} + {}^{234}_{90}\text{Th}$

e. $^{3}_{1}\text{H} \xrightarrow{b} {}^{0}_{-1}\text{e} + {}^{3}_{2}\text{He}$

99. Three common methods for detecting radiation include photographic films, Geiger counters, and scintillation counters.

101. A rad is the unit used to express the amount of energy associated with radiation. One rad is equal to 1×10^{-2} J of energy being absorbed by 1 kg of exposed tissue.

103. An LD_{50} is the lethal dose for 50% of the population. In terms of radiation, the LD_{50} is the amount of radiation where fifty percent of the exposed population would perish.

105. Fission occurs when a large nucleus splits into smaller nuclei. Uranium-235 is an example of an isotope that undergoes fission.

107. For fusion to occur, light nuclei and very high temperatures are required.

109. If a mass of uranium-235 is supercritical, an uncontrollable chain reaction will occur that will release an enormous amount of energy.

111. Nuclear reactions and nuclear energy are used as an alternative fuel source, in food processing to improve shelf life, and in medicine for the diagnosis and treatment of medical conditions.

113. The atomic mass for chlorine from these data is 35.470382. This is within two-hundredths of what is listed on the periodic table.

115. The volume of the nucleus is 4.1888×10^{-39}cm. The volume of the atom is 4.1888×10^{-24}. Remembering that a percentage is equal to the part divided by the total times one hundred, the percentage of the atoms volume that is occupied by the nucleus is 0.0000000000001% or 1×10^{-13}%.

Sample Test:

1. Two atoms of the same element cannot be different in the number of _____________?
   a. protons b. neutrons c. electrons d. the atoms must be the same

2. What happens during a chemical reaction?
   a. atoms are destroyed
   b. atoms are created
   c. atoms are neither created nor destroyed
   d. atoms are both created and destroyed

3. Nucleons are the subatomic particles found in the core of an atom. These subatomic particles are
   a. protons b. electrons c. neutrons d. more than one correct answer

4. The periodic table is arranged in order of increasing atomic number. The atomic number represents the number of ____________ in an atom?
   a. protons b. electrons c. neutrons d. nucleons

5. Give the number of protons, neutrons, and electrons for a neutral atom of gold, respectively?
   a. 47, 60, 47 b. 79, 79, 118 c. 79, 118, 79 d. 31, 39, 31

6. Fill in the table below.

Element Name	Element Symbol	Atomic Number	Mass Number	Number of Protons and Electrons	Number of Neutrons
		12			
	O				
				26	
Calcium					

7. Different isotopes of an element have different numbers of
   a. protons b. electrons c. neutrons d. protons and neutrons

8. The element phosphorous has an atomic number of 15. How many shells, valence electrons, and core electrons would be found for this element, respectively?
   a. 3, 3, 12 b. 3, 5, 10 c. 4, 12, 3 d. 4, 10, 5

9. How many electrons are in each shell of an atom of potassium-19?

a. 8, 8, 3 b. 2, 8, 8, 1 c. 10, 8, 1 d. 19

10. For an atom of sulfur-16, which shell has the highest energy?
    a. First b. Second c. Third d. All the shells have the same energy

11. Elements emit light when
    a. electrons move to higher shells
    b. electrons move to lower shells
    c. core electrons move to the valence shell
    d. more than one correct answer

12. How would you expect the light emitted from helium to compare to the light emitted by hydrogen.
    a. Hydrogen would have more lines of light emitted.
    b. Helium would have more lines of light emitted.
    c. Hydrogen and helium would have the same number of lines of light emitted.
    d. Not enough information.

13. Identify the element in Group 4A and period 5.
    a. As b. Sn c. Zr d. V

14. Which of the following represents a metal and nonmetal combination?
    a. C and O b. P and Cl c. K and Br d. B and Cl

15. Which elements are electrical insulators?
    a. alkali metals b. metalloids c. nonmetals d. alkaline earth metals

16. How many valence electrons do each of the elements in group 7A have?
    a. 2 b. 5 c. 7 d. it depends on the element

17. An element in the third period will have what different subshells present?
    a. s b. s, p c. s, p, d d. s, p, d, f

18. What is the electron configuration of titanium?
    a. $1s^22s^22p^63s^23p^64s^4$ b. $1s^22s^22p^63s^23p^63d^2$ c. $1s^22s^22p^63s^23p^64s^2$ d. $1s^22s^22p^63s^23p^64s^23d^2$

19. How many orbitals are in a d subshell?
    a. 1 b. 3 c. 5 d. 7

20. Nitrogen-13, which is used to image organs in the body, has a half-life of 10 minutes. If you are injected with a 50 cc dosage of nitrogen-13, how much of the radioactive isotope would remain in your system after 30 minutes?
    a. 50 cc b. 25 cc c. 12.5 cc d. 6.25 cc

21. The emission of a beta particle from plutonium-242 (Pu) would yield what element?
    a. Np b. Am c. Cm d. U

22. If Superman holds onto a block of krypton, a gamma emitter, what is it that causes him to lose his power?
    a. an electron
    b. a molecule of helium
    c. a high-energy particle with no appreciable mass
    d. Lex Luther

23. What element is formed upon alpha emission from an atom of Radon (Rn)?
    a. Ra b. Po c. Pb d. Th

24. Position Emission Tomography detects tumors by hitting positrons into electrons. What types of particles are detected by this method?
    a. alpha particles b. beta particles c. gamma particles d. neutrons

25. Name three units used to express radiation?

26. What effect will excessive exposure to radiation have?
    a. you will glow in the dark b. you will sink in water c. damage to proteins and genetic material d. none

27. In the movie, "Back to the Future", the professor used radioactive plutonium-242 to fuel the nuclear reactor in his DeLorean time machine. Assuming that he used the process of nuclear transformation and the particle used for bombardment was a neutron, after fusion and beta decay, what element would remain?
    a. americium-243 b. plutonium-243 c. neptunium-238 d. uranium-243

28. Nuclear power plants attempt to collect a fraction of the energy released by what type of nuclear reaction?
    a. a supercritical fusion chain reaction
    b. a critical fusion chain reaction
    c. a subcritical fission chain reaction
    d. a critical fission chain reaction

29. If Uranium-235 is the primary source of energy for nuclear weapons, why is strontium-90 getting into the food chain through radioactive fallout?
    a. fusion of uranium yields strontium
    b. fission of uranium yields strontium
    c. it was a coincidence that strontium was detected
    d. uranium-235 is not radioactive

30. An intense beam of gamma radiation from cobalt-60 is focused on tumors to destroy them. What happens to cobalt-60 when it emits gamma radiation?
    a. it turns into nickel-60
    b. it turns into copper-64
    c. it blows apart by fission into various different atoms
    d. it remains cobalt-60

Answers to Sample Test:

1. a
2. c
3. d
4. a
5. c
6.

Element Name	Element Symbol	Atomic Number	Mass Number	Number of Protons and Electrons	Number of Neutrons
Magnesium	Mg	12	24	12	12
Oxygen	O	8	16	8	8
Iron	Fe	26	56	26	30
Calcium	Ca	20	40	20	20

7. c
8. b
9. b
10. c
11. b
12. b
13. b
14. c
15. c
16. c
17. b
18. d
19. c
20. c
21. b
22. c
23. b

24. c
25. The three units commonly used to express radiation are curies, rads, and rems.
26. c
27. a
28. d
29. b
30. d

Chapter 3: Chemical Bonding

Chapter Objectives:

1. To use Lewis structures for the representation of atoms, molecules, and monatomic ions.
2. To understand how ions are formed in order to predict the charge on monatomic ions.
3. To use the knowledge of ion formation to understand ionic compounds and ionic bonding.
4. To understand the sharing of electrons that results in the formation of covalent bonds.
5. To use the periodic table to predict the type of bond that will form between two atoms.
6. To be able to visualize the three-dimensional shapes of molecules.
7. To predict the polarity of a bond and a molecule.

Key Terms: The key terms are listed in the order in which they are encountered in the Chapter.

Unreactive. An atom that has very little tendency to combine with other atoms.

Reactive. An atom that will readily combine with other atoms.

Octet rule. Atoms will gain, lose, or share electrons to obtain eight electrons in their valence shell with the exception of hydrogen and helium.

Lewis structure. The electron dot structure or Lewis structure uses the symbol of an element to represent the core electrons and puts dots around the symbol to represent the number of valence electrons for that element.

Ion. An atom or group of atoms that carries an electrical charge.

Cation. An atom that loses one or more electrons becomes positively charged in the process.

Anion. An atom that becomes negatively charged through the addition of electrons.

Monatomic ion. An ion or charged particle that contains only one atom.

Polyatomic ion. An ion that contains more than one atom.

Ionic bond. The force of attraction between oppositely charged bonds.

Ionic compound. A compound that consists of ions held together by ionic bonds.

Electrolytes. Compounds that conduct electricity.

Covalent bond. The force of attraction between two atoms that is the result of a sharing of electrons.

Molecule. A group of atoms joined together by covalent bonds.

Molecular compound. A compound made up of molecules that do not contain ions and are typically nonelectrolytes.

Single bond. This is a single covalent bond that involves the sharing of a single pair of electrons.

Diatomic molecule. A molecule that contains only two atoms.

Double bond. A covalent bond involving two pairs of shared electrons.

Triple bond. A covalent bond involving three pairs of shared electrons.

Partial positive (δ^+). The atom at the end of a polar covalent bond that takes on a slight positive character due to an unequal sharing of electrons.

Partial negative (δ^-). The atom at the end of a polar covalent bond that takes on a slight negative character due to an unequal sharing of electrons.

Dipole. The separation of partial charges associated with the unequal sharing of electrons.

Polar covalent bond. A polar bond that joins two atoms of opposite partial charges. A dipole exists across the bond.

Nonpolar covalent bond. Two atoms that equally share the electrons that bind them together.

Electronegativity. The tendency of bonding atoms to attract electrons to themselves.

Linear. The arrangement in space of atoms that places their valence electrons as far apart as possible leaving the atoms oriented 180° apart from one another.

Bent. The arrangement in space of atoms that places their valence electrons as far apart as possible leaving the atoms oriented at an angle of 104.5° apart from one another.

Pyramidal. The arrangement in space of three atoms about a central atom that places their valence electrons as far apart as possible leaving the atoms oriented at an angle of 109.5° apart from one another.

Tetrahedral. The arrangement in space of four atoms about a central atom that places their valence electrons as far apart as possible leaving the atoms oriented at an angle of 109.5° apart from one another.

Lone pair electrons. Lone pair or unshared electrons are electrons that are left over or nonbonding in the Lewis dot structures of molecules that contribute to the overall geometry of the molecule.

Chapter 3 Objective Details:

a. Lewis Structures. Lewis dot structures are based on the propensity for every atom to be stable. The most stable electronic configuration for an element is to have eight electrons in the valence shell. Atoms will gain, lose, or share electrons to achieve valence electrons or an octet. This is known as the octet rule. The exception to the rule includes atoms of the first shell. The first shell can accommodate two electrons in the s subshell. Helium is a member of the noble gases because the valence shell is filled, but it only requires two electrons not eight.

In a Lewis dot structure, the element symbol is written and then dots are used to represent the valence electrons. For the elements in groups 1A through 8A, the group number corresponds to the number of valence electrons. For the elements that are not in groups 1A through 8A, it is recommended that the student write the entire electron configuration for the atom and determine the valence electrons based on the highest shell.

Example: Draw the Lewis dot structure for tin.

Solution:

1.) The symbol for tin is Sn.
2.) Tin has the atomic number 50. Do we really have to write out the entire electron configuration for tin? No! Since tin is in group 4A, we can conclude that it will have four valence shell electrons.
3.) We start by writing the symbol for tin to represent all of the core electrons.
4.) The last step is to include the valence electrons around the perimeter of the symbol.

$$\cdot\overset{\cdot}{\underset{\cdot}{\mathrm{Sn}}}\cdot$$

5.) It should be noted that each valence electron is drawn one per side of the atom. This is done intentionally in order to allow for an octet of electrons when the atom reacts with another atom. In the case of tin, there are four electrons that are situated one per side of the symbol. This means that tin has four valence electrons and needs four more to complete its octet.

b. Ions. Ions are charged particles formed by the gain or loss of electrons. The two types of ions are monatomic and polyatomic. The difference between them is that monatomic ions contain only one atom, and polyatomic ions contain more than one atom. Ions are formed because the elements in groups 1A through 3A, the metals, have a tendency to lose electrons to satisfy the octet rule. Elements in groups 1A, 2A, and 3A will lose 1, 2, and 3 electrons, respectively to acquire charges of +1, +2, and +3, respectively. This is true for all of the metals in groups 1A through 3A with the exception of boron, which will be addressed later. Since metals lose electrons, it makes sense that the nonmetals of groups 5A, 6A, and 7A will gain 3, 2, and 1 electron, respectively to form charges of –3, -2, and –1, respectively. It should be remembered that the charge on an electron is a negative one. To lose an electron is to gain a positive charge and to gain an electron is to gain a negative charge. When a metal and a nonmetal come into contact, the metal will donate enough electrons to satisfy its octet while the nonmetal will accept enough electrons to complete its octet.

Example: Lets take the case of sodium interacting with chlorine to make table salt.

1. Sodium is a metal in group 1A. All of the elements in group 1A have one valence electron. In order to complete the octet, sodium will want to give that electron away to become more stable.
2. Chlorine is a nonmetal in group 7A. All of the elements in group 7A have 7 valence electrons. In order to complete the octet, chlorine will want to gain an electron.
3. When sodium and chlorine come into contact, sodium gives an electron to chlorine thus acquiring a positive one charge. Chlorine, which accepts the electron, thus gains a negative one charge.
4. The result of this is that both atoms have become ions and in so doing have stabilized one another.
5. The Lewis dot structure for the transfer of an electron between sodium and chlorine is shown in the equation below. Notice that chlorine only has one available spot to accept an electron and sodium only has one electron to give. Also notice that the charges on the ions are drawn toward the top right of the elements symbol.

$$\text{Na}\cdot + \cdot\ddot{\underset{\cdot\cdot}{\text{Cl}}}\colon \longrightarrow \text{Na}^{+} + \left[\colon\ddot{\underset{\cdot\cdot}{\text{Cl}}}\colon\right]^{-}$$

c. Ionic bonds and ionic compounds. When an anion is in the presence of a cation, an ionic bond is the result. Students often confuse the term ionic bond. In an ionic compound, it is really a force of attraction that holds the oppositely charged atoms together. In later discussions of covalent bonds, it should be remembered that ionic bonds are really ionic interactions. Ionic compounds will in general conduct electricity and are considered to be electrolytes.

d. Covalent bonding. Covalent bonds represent the sharing of electrons between atoms. In ionic bonding, electrons are not shared they are transferred. Covalent bonds represent a shared pair of electrons. There can exist one, two, or three pairs of shared electrons that are termed single, double, and triple covalent bonds, respectively. When two or more atoms share their electrons in the form of covalent bonds, they are called molecules. Since not all atoms have the same affinity for electrons, covalent bonds often have dipoles associated with them. A dipole is a difference in charge between two atoms across a covalent bond. If one atom has a higher affinity for an electron than the atom it is bound to, then the atom will have a partial negative charge. The opposite is also true. It should be remembered that metals and nonmetals mix to make ionic compounds and it is also true that nonmetals will combine with nonmetals to form covalent compounds.

e. Electronegativity: A measure of the relative attraction that an atom has for the shared electrons in a bond.

What to remember:

a. Fluorine: E=4.0
b. Hydrogen: E=2.1
c. $\Delta E > 1.8$ = ionic
d. $\Delta E < 1.8$ = covalent
e. $\Delta E = 0.0$ = nonpolar covalent

Nonpolar covalent bond: A covalent bond in which there is equal sharing of electrons ($\Delta E = 0.0$).

Polar covalent bond: A covalent bond in which there is unequal sharing of electrons ($\Delta E < 1.8$).
δ^- = fractional negative charge
δ^+ = fractional positive charge
Ionic bond: Electron transfer that occurs on account of an electronegativity difference between atoms of more than 2.1 ($\Delta E > 1.8$ = ionic).

The periodic table is arranged in such a way that there is an electronegativity trend towards the top right where fluorine is located. Fluorine is the most electronegative element on the periodic table with an electronegativity of 4.0. Electronegativities of the elements increase across the periodic table from left to right and up the table from bottom to top. The chart below gives electronegativities for many of the common elements and attempts to show the electronegativity trend that is observed.

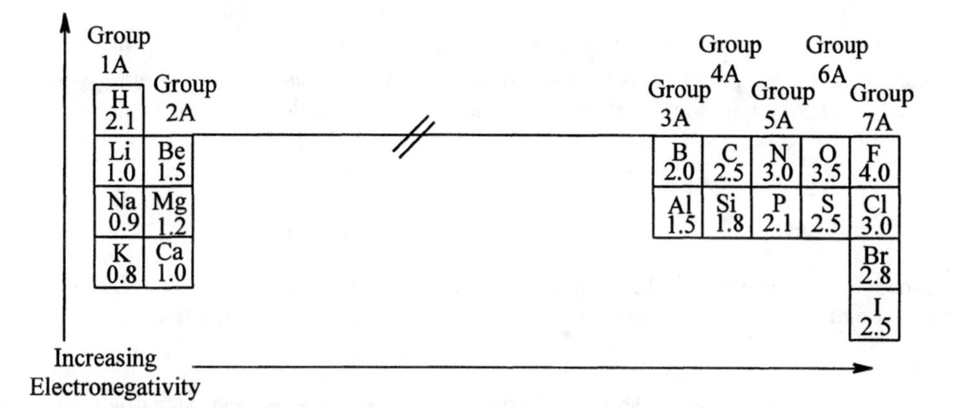

f. Shapes of Molecules. Molecules can exist in a linear, bent, pyramidal, or tetrahedral geometric shape depending on the number of electron pairs that exist. The first step in predicting the shape of a molecule is to draw the Lewis dot structure for each of the atoms. Based on your knowledge of covalent bonding, you can determine how many bonds an element will form. A general rule of thumb is called HONC. HONC stands for the number of bonds that will form for the elements hydrogen (1), oxygen (2), nitrogen (3), and carbon (4). If you know that hydrogen will form one bond, oxygen will form two bonds, nitrogen will form three bonds, and carbon will form four bonds, then you can use your knowledge of the periodic table to predict the number of bonds that will form for other elements that have the same number of valence electrons. For example, carbon is in group 4A on the periodic table. It has four valence electrons and will generally form four bonds. That means that the other members of group 4A will also have the same tendency to form four bonds.

How does the number of bonds and Lewis structure help to predict the shape of the molecule? One way to think of this is that groups of electrons get as far away from each other as possible. In the presence of a repulsive force, like lone pair electrons, bonds will be bent to get the electrons as far away from each other as possible. Let's take a look at an example.

Example: Predict the shape of a molecule of ammonia (NH_3).

Solution:

1.) Draw the Lewis dot structure.

```
  ..
H:N:H
  ..
  H
```

2.) Determine the number of bonds that form between the atoms.

Remember HONC! Nitrogen will often form three bonds and hydrogen will form one bond. If there are three hydrogen atoms and only one nitrogen, it makes sense that there will be three single bonds between nitrogen and hydrogen.

3.) Are there any lone pair electrons? If so, they take the place of an additional bond in terms of repulsive force.

If you make the three bonds between hydrogen and nitrogen, you will notice that the nitrogen has two electrons that are not bonding to anything. These lone pair electrons will repel the hydrogen atoms in the same manner as another atom.

4.) How many atoms are there and what is their shape?

There are three hydrogen atoms located around the perimeter of the nitrogen atom. The tendency is to form a tetrahedral shape, but there are only three peripheral atoms. Because there are only three peripheral atoms, the shape of the molecule is pyramidal.

g. Molecular Polarity. A molecule consisting of only nonpolar bonds will be nonpolar regardless of its shape. A molecule that contains a polar bond or bonds, can be either polar or nonpolar. The molecular geometry and the bond dipoles must be known to predict the polarity of the molecule. If there is an unequal dipole and a nonsymmetrical shape to the molecule, it will be polar. If there is a dipole of equal and opposite magnitude, the molecule will be nonpolar.

Solutions to Odd Numbered Problems:

1. All of the noble gases except helium have eight valence electrons. Helium only has one shell that contains only the s subshell so it can accommodate no more than two electrons. It is still a noble gas because the outer most shell is completely filled.

3. Elements that are not noble gases will gain, lose, or share electrons to achieve a noble-gas electron configuration.

5. a. Lithium is in the second period and will therefore have two shells. The first shell will contain two electrons to fill the s subshell. The second shell will contain one electron to give a total of three electrons for lithium. Since there is only one electron in the outermost or second shell, there is only one valence electron. A quick check to confirm that lithium has only one valence electron is to look at the periodic table. Lithium is in group 1A and therefor is only expected to have one valence electron.
   b. Beryllium has one more electron than lithium, so it has two valence electrons. Beryllium is in group 2A and is therefor expected to only have two valence electrons. In terms of electron shells, beryllium has four total electrons that fill the first shell and the s subshell of the second shell.
   c. Potassium is in group 1A and is expected to have only one valence electron. The electron configuration for potassium is $1s^22s^22p^63s^23p^64s^1$. The highest shell is the fourth and there is only one electron in that shell. All evidence is consistent with the fact that potassium has only one valence electron.

7. a. 1
   b. 6
   c. 7
   d. 8 Argon satisfies the octet rule.
   e. 2
   f. 3

9. In a Lewis structure, the element symbol represents the core electrons for an atom.

11.

a. Li· b. :Ö· c. ·Ċ· d. Ȧl·

13.

a. K· b.. S: c.·Cl. d.:Ar: e. Ċa· f. Ḃ·

15. An ion is an atom or group of atoms that carries an electrical charge. A sodium atom is different than a sodium ion in that the sodium ion has donated an electron to acquire a positive charge.

17. An anion is a negatively charged ion. All of the elements in groups 5A, 6A, and 7A will form anions in order to complete their octet. Two examples of anions are N^{3-} and O^{2-}.

19. The charge on an ion is represented as a superscript following the elemental symbol. If the charge is other than plus or minus one, the number representing the magnitude of the charge will precede the plus or minus sign.

21. a. S^{2-} is a anion.
    b. Ag^{+} is an cation.
    c. Cr^{3+} is a cation.

23. A monatomic ion contains only one atom while a polyatomic ion contains more than one atom. The cyanide ion is made by the interaction of carbon and nitrogen. Since there are more than one element involved in the ion, cyanide would be considered polyatomic.

25. When a nonmetal is involved in an electron transfer reaction, it will accept electrons to form an anion. Bromine is a member of group 7A on the periodic table. This means that bromine has seven valence electrons and will gain one electron to complete its octet. Upon the addition of an electron, the anion Br^{-} will be formed.

27. a. Oxygen is a nonmetal in group 6A and would gain two electrons.
    b. Magnesium is a metal in group 2A and would lose two electrons.
    c. Sodium is a metal in group 1A and would lose one electron.
    d. Tellurium is a nonmetal in group 6A and would gain two electrons.

29. a. –2
    b. +2
    c. +1
    d. -2

31.

a. $[:\ddot{\underset{..}{O}}:]^{2-}$ b. Mg^{2+} c. Na^{+} d. $[:\ddot{\underset{..}{Te}}:]^{2-}$

33.

a. $2\ Li\cdot + \cdot\ddot{\underset{..}{S}}\cdot \longrightarrow 2\ Li^{+} + [:\ddot{\underset{..}{S}}:]^{2-}$

b. $\cdot Ca\cdot + 2\ \cdot\ddot{\underset{..}{Br}}: \longrightarrow Ca^{2+} + 2\ [:\ddot{\underset{..}{Br}}:]^{-}$

c. $\cdot\underset{\cdot}{Al}\cdot + 3\ \cdot\ddot{\underset{..}{Cl}}: \longrightarrow Al^{3+} + 3\ [:\ddot{\underset{..}{Cl}}:]^{-}$

35. Oxygen needs two electrons to complete its octet and magnesium needs to lose two electrons to complete its octet. The ratio of the elements will be one to one. The oxygen atom will receive two electrons from the metal magnesium to form an anion with a charge of negative two. The magnesium will donate two electrons to the nonmetal oxygen to from a cation with a charge of positive two.

37. An ionic bond is the force of attraction between oppositely charged ions.

39. Ionic compounds contain ions bound by ionic bonds. Thus to determine if two elements will form an ionic compound, we should look for a metal interacting with a nonmetal. In this case, zinc and chlorine (b) is the only case where a metal/nonmetal interaction is occurring.

41. An ionic compound is any compound that consists of ions held together by ionic bonds. Ionic compounds typically form crystalline solids. By looking to see if the compounds is crystalline will be an indication of whether it is likely of being an ionic compound.

43. Two nonmetals will not form an ionic compound when mixed. Sulfur and oxygen are both in group 6A and will both be reluctant to give up electrons. Nonmetals take electrons and will only form ionic compounds with metals that will donate electrons.

45. Electrolytes are ions or compounds that conduct electricity. Some examples of electrolytes in the body include ions of calcium, potassium, magnesium, sodium, chloride, and bicarbonate.

47. A molecule is a group of atoms joined together by covalent bonds.

49. a. Lead is considered to be a metal and oxygen is a nonmetal, so they will combine to form an ionic compound.
b. Sulfur and fluorine are both nonmetals and will combine to form a covalent compound or molecule.
c. Phosphorus and hydrogen will combine to form a covalent compound.

51. nonelectrolyte

53. A single bond consists of a shared pair of electrons. A double bond consists of two shared pairs of electrons. A triple bond consists of three shared pairs of electrons.

55. A diatomic molecule is a molecule that contains only two atoms.

57. A dipole is a separation of charge across a covalent bond.

59. Fluorine attracts electrons better than does phosphorous, so fluorine will have more negative character and a partial negative charge. If fluorine is partially negative, then phosphorous must be partially positive.

$$\overset{\longleftarrow\!\!+}{\underset{\delta^-\quad\delta^+}{F\text{—}P}}$$

61. a. Silicon and hydrogen are both nonmetals and will thus form covalent bonds when in contact with one another.
b. Lithium is a metal and tellurium is a nonmetal. When combined, they will form an ionic bond.
c. Nitrogen and hydrogen are both nonmetals and will form covalent bonds.
d. Cadmium is a metal and fluorine is a nonmetal. When joined, they will form an ionic bond.

63. The elements in the top right hand side of the periodic table would have the highest electronegativity values. The elements in the bottom left of the periodic table would have the lowest electronegativity values.

65. a. The difference in electronegativity between sodium (E=1.0) and bromine (E=2.8) is 1.8. The bond between the two will be ionic.
b. The difference in electronegativity between hydrogen (E=2.1) and bromine (E=2.8) is 0.7. The bond between them will be polar covalent.
c. There is no difference in electronegativity between two atoms of bromine. The bond that will form between two elements of bromine will be nonpolar covalent or just covalent.
d. The difference in electronegativity between carbon (E=2.5) and oxygen (E=3.5) is 1.0. The bond that will form between them will be polar covalent.

67. Repulsions between electrons are minimized by groups of electrons getting as far away from one another as possible.

69. Nitrogen and phosphorus are in the same group on the periodic table. Elements from the same groups will generally form molecules with the same geometry. PH_3 will have a pyramidal shape.

71. A nonpolar molecule can contain polar bonds that are offset by other polar bonds depending on the geometry of the molecule. Just because the molecule is nonpolar does not mean that the bonds are nonpolar.

73. a. CS_2 is nonpolar because the polar bonds are of equal and opposite magnitude.
    b. PCl_3 is polar because the polar bonds are not of equal and opposite magnitude.
    c. HF is polar because there is a distinct difference in electronegativity between hydrogen and fluorine.

75. Like dissolves like means that nonpolar molecules will be soluble in nonpolar solvents, and polar molecules will be soluble in polar solvents.

77. For boron to position three fluorine atoms as far apart as possible, the fluorine atoms would have to be at 120° angles from one another in the same plane as boron.

Sample Test:

1. How many valence electrons are found in the alkali metals, alkaline earth metals, halogens, and noble gases, respectively?
   a. 2, 1, 8, 7 b. 1, 2, 8, 7 c. 1, 2, 7, 8 d. not enough information

2. If an element is in group 5A, how many valence electrons will it have?
   a. 3 b. 5 c. 7 d. 9

3. How many valence electrons do each of the following elements have?
   a. Cl b. Mg c. N d. Xe

4. What is the correct Lewis structure for sulfur?

   a. $\cdot\overset{\cdot\cdot}{\underset{\cdot}{S}}\cdot$ b. $:\overset{\cdot\cdot}{\underset{\cdot\cdot}{S}}:$ c. $\cdot\overset{\cdot}{\underset{\cdot}{S}}\cdot$ d. $:\overset{\cdot}{\underset{\cdot}{S}}:$

5. What charge would you expect strontium to acquire upon ionization?
   a. +1 b. +2 c. -1 d. -2

6. How many atoms of magnesium will be required to react with two atoms of aluminum?
   a. 1 b. 2 c. 3 d. 6

7. What will happen to sulfur if it reacts with two atoms of potassium?
   a. lose one electron b. lose two electrons c. gain one electron d. gain two electrons

8. Bananas are a good source of potassium ions. Potassium ions along with sodium and chloride ions are involved in the electrical potential of cell membranes. How do potassium ions aid in maintaining the electrical potential across cell membranes?
   a. potassium forms a negatively charged ion that will conduct electricity
   b. potassium forms a positively charged ion that will conduct electricity
   c. potassium forms an ionic compound with sodium that conducts electricity
   d. potassium is a nonmetal that does not conduct electricity

9. Lead and mercury both form ions that interact with the sulfur atoms of proteins in the human body. Ionic bonds form between these "heavy metals" and the sulfur atom. Based on your knowledge of the periodic table, what charge will be associated with the lead and mercury ions that interact with sulfur?
   a. -1 b. -2 c. +1 d. +2

10. Sodium chloride is an electrolyte while carbon dioxide is a nonelectrolyte. Why?
    a. carbon dioxide forms an ionic compound
    b. sodium chloride forms a covalent compound
    c. carbon dioxide forms a covalent compound
    d. whether the compound is ionic or covalent doesn't have any effect on the electrolytic properties

11. Based on your knowledge of chemistry, identify the type of bond in each of the following molecules.

    a. $\underset{\delta^-}{F}-\underset{\delta^+}{H}$ b. $CaCl_2$ c. H_2 d. $C\equiv O$

12. What's the difference between a molecule and a compound?
    a. A compound represents atoms that are joined by covalent bonds.
    b. A molecule represents atoms that are joined together by covalent bonds.
    c. Compounds are nonelectrolytes.
    d. Molecules are electrolytes.

13. Why does the combination of a metal and a nonmetal form an ionic bond?
    a. The metal will share its valence electrons with the nonmetal.
    b. The nonmetal will share its valence electrons with the metal.
    c. The metal will give its valence electrons to the nonmetal.
    d. The nonmetal will give its valence electrons to the metal.

14. Predict whether the bond that forms between the following combinations of elements will be ionic, polar covalent, or nonpolar covalent?
    a. C and O b. P and Cl c. K and Br d. B and Cl

15. Why is electronegativity such an important concept?
    a. Electronegativity helps to predict whether a bond between two atoms will be nonpolar covalent.
    b. Electronegativity helps to predict whether a bond between two atoms will be polar covalent.
    c. Electronegativity helps to predict whether a bond between two atoms will be ionic.
    d. More than one correct answer.

16. What is the difference in electronegativity between sodium and chlorine? What type of bond will form between an atom of sodium and an atom of chlorine?
    a. The difference in electronegativity is more than 1.8, so the bond will be polar covalent.
    b. The difference in electronegativity is more than 1.8, so the bond will be ionic.
    c. The difference in electronegativity is more than 1.8, so the bond will be nonpolar covalent.
    d. The difference in electronegativity is less than 1.8, so the bond will be polar covalent.

17. Why is water bent and not linear?
    a. It is linear.
    b. It is bent because it has two hydrogen atoms.
    c. It is bent because oxygen has two lone pair electrons that repel each other and the hydrogen atoms.
    d. It is really pyramidal.

18. What will be the geometry of SiH_4?
    a. linear b. bent c. pyramidal d. tetrahedral

19. What will be the geometry of carbon dioxide (CO_2)?
    a. linear b. bent c. pyramidal d. tetrahedral

20. The fact that oil does not dissolve in water means that …
    a. oil is nonpolar   b. oil is polar   c. water is nonpolar   d. it doesn't mean anything

Answers to Sample Test:

1. c
2. b
3. a. 7
   b. 2
   c. 5
   d. 8
4. d
5. b
6. c
7. d
8. b
9. d
10. c
11. a. polar covalent
    b. ionic
    c. nonpolar covalent
    d. polar covalent
12. b
13. c
14. a. polar covalent
    b. polar covalent
    c. ionic
    d. polar covalent
15. d
16. b
17. c
18. d
19. a
20. a

Chapter 4: The Language of Chemistry

Chapter Objectives:

1. To use chemical formulas to determine the components of compounds.
2. To write and interpret formulas for ionic compounds.
3. To write and interpret formulas for covalent compounds.
4. To express and interpret chemical equations.
5. To understand and use moles to express chemical equations.

Key Terms: The key terms are listed in the order in which they are encountered in the Chapter.

Formula. A chemical formula shows the composition or structure of the smallest whole-number ratio of ions present in a compound.

Formula Unit. The smallest whole-number ratio of ions present in a compound.

Molecular Formula. Displays the composition of a molecule by giving the symbol of each element and then the number of atoms of each element as a subscripted number.

Structural Formula. Displays the connectivity of atoms to one another in a molecule.

IUPAC. The International Union of Pure and Applied Chemistry that has developed rules that link chemical names to chemical composition and structure.

Binary ionic compounds. Ionic compounds containing ions of only two elements where the first element listed is always the metal and the second element listed is always the nonmetal.

Binary covalent compound. A compound that is comprised of atoms of two nonmetals bound together by covalent bonds.

Reactants. Substances that react to form new substances.

Products. The resulting materials from a chemical reaction.

Chemical reaction. The process by which reactants are converted to products.

Chemical equation. A concise symbolic representation of a chemical reaction.

Aqueous (aq). Water; when a substance is dissolved in water, the medium is described as aqueous.

Coefficients. Numbers used to indicate the relative number of molecules, atoms, or ions that participate in the reaction.

Law of Conservation of Matter. Matter is neither created nor destroyed during the course of a chemical reaction.

Balanced chemical equation. A chemical equation is balanced when there are the same number of atoms of each element on both sides of the arrow separating reactants from products.

Stoichiometry. The quantitative relationship between reactants and products in a chemical reaction.

Mole. A counting unit that represents 6.022×10^{23} of anything.

Avogadro's number. The number 6.022×10^{23} that has been termed the mole.

Atomic mass. The mass of one atom of an element expressed in amu, or the mass of one mole of an element expressed in grams.

Molar mass. The atomic mass of an element expressed in grams.

Formula mass. The mass of an ionic compound is the total mass of the ions in the formula of that compound.

Molecular mass. The total mass of a molecular compound is the total mass (in amu) of the atoms in a molecule of the compound.

Chapter 4 Objective Details:

a. Formulas. Ionic compounds typically form crystalline solids that are structurally complex. The chemical formula for an ionic compound will give the formula unit or smallest whole number ratio of ions in the compound. To determine the number of ions of one element that are necessary to bind to another ion, remember to cancel the charges. If one ion has a positive two charge and the other ion has a negative three charge, swap the charges for the ions for the number of ions in the compound.

Take calcium fluoride as an example. This compound is a mixture of calcium and fluorine, but what is the ratio or chemical formula for the compound?

Calcium is a metal in group 2A and will ionize (lose two electrons) to form a cation with a positive two charge. Fluorine is a halogen in group 7A and will ionize (gain an electron) to form an anion with a negative one charge. The charge on calcium is +2, so take that number as the number of fluorine ions necessary for the compound. The charge on fluorine is –1, so take that number as the number of calcium ions necessary for the compound. Swapping charges will give the correct chemical formula of CaF_2. The symbols for the elements are given with the metal being listed first and the nonmetal second. Notice the two fluorine atoms are designated by the subscripted number.

Molecular formulas are the same as ionic formulas in the sense that they show the symbol for each element in a molecule and the number of atoms of each element as a subscripted number.

Structural formulas give a diagrammatic representation of how atoms are connected to one another in a molecule.

b. Naming ionic compounds. Ionic compounds are named with the metal first and the nonmetal second. The metal ion will always form a cation and the nonmetal will always form an anion. The name of the metal is the same as the name of the element, but the name of the nonmetal is the truncated element name followed by the suffix *–ide*. A general rule of thumb is that all of the elements or metals in group 1A, 2A, and 3A will have charges of +1, +2, and +3, respectively. The elements or nonmetals in group 7A, 6A, and 5A will have charges of –1, -2, and –3, respectively. It can be assumed that the metals found in the center of the periodic table may have more than one cationic form. For these elements, the charge of the cation must be indicated by a Roman numeral. The Roman numerals most commonly encountered are I, II, III that represent ionic charges of +1, +2, and +3, respectively. Lets look as some examples.

Example: Determine the chemical formula for potassium sulfide.

Solution:

1.) Where are the elements on the periodic table?
    Potassium is in group 1A and sulfur is in group 6A.
2.) What charge will each ion form?
    Potassium will form a cation with a +1 charge.
    Sulfur will form an anion with a –2 charge.
3.) How many of each atom will need to be present for the charge on the molecule to be neutral?
    Two atoms of potassium will cancel the charge on one atom of sulfur.
4.) Using the symbols for the elements, write the chemical formula for potassium sulfide.
    There will be two potassium atoms for every one sulfur atom so the formula will be K_2S.

Example: Determine the chemical name for the compound Cu_2O?

1.) Where are the elements on the periodic table?
Copper is in group 1B, and oxygen is in group 6A.
2.) What charge will each ion form?
Copper is in a part of the periodic table where multiple charges are possible.
Oxygen is in group 6A, so it can be assumed that the charge on the ion will be –2.
3.) What will be the charge on copper for the charge on the compound to be neutral?
Since the charge on oxygen is –2, copper must have a charge of +1, for each of the two atoms, in order for the compound to be neutral in charge.
4.) Using the above information, what is the name of the ionic compound with the formula Cu_2O?
Since copper is capable of having more than one cationic form, the charge for the cation in this compound must be specified with a Roman numeral. The charge for copper is +1, so the Roman numeral for the cation is I. Oxygen is the nonmetal, so the –ygen is dropped and –ide is added to indicate that the oxygen is part of a compound. The name for Cu_2O is copper (I) oxide.

Polyatomic ions are named in exactly the same way as regular ions. The difference with polyatomic ions, is the fact that the ion contains more than one atom.

c. Naming covalent compounds. Covalent compounds are by definition combinations of nonmetals that are linked together by covalent bonds. When naming these compounds, the less electronegative of the two elements comes first in the name. Remembering the periodic trend for electronegativity indicates that the element to be named first will be either further to the left or further down on the periodic table than the element being named second. The name of the second element is truncated and the suffix *–ide* is added. Unlike ionic compounds, covalent compounds use the prefixes *mono, di-, tri-, tetra-, penta-,* and *hexa-* to indicate that there are one, two, three, four, five, and six atoms of an element present, respectively.

Example: Take disilicon hexafluoride, what can you tell from the name and what is the formula for this compound?

Solution:

1.) The prefix *di-* indicates that there are two atoms of silicon present in the formula.
2.) The prefix *hexa-* indicates that there are six atoms of fluorine present in the formula.
3.) The *–ine* is dropped off of fluorine and *–ide* is added.
4.) The formula for disilicon hexafluoride is Si_2F_6.

d. Chemical equations. Chemical equations represent the conversion of reactants into products. The reactants are always shown on the left followed by an arrow and then the products are shown to the right. The law of conservation of energy indicates that the atoms that react must be present as the same atoms in the product. The arrangement of atoms will change, but the number of atoms of each element will stay the same. It is for this reason that chemical equations must be balanced. To balance a chemical equation, make the number of reacting elements equal to the number of elements produced. One more thing to keep into account is that a parenthesis containing an italicized *l*, *s*, *g*, or *aq* indicates the state of liquid, solid, gas, or aqueous respectively.

To balance a chemical equation, start with an element that only appears once on each side of the equation. Balance those elements first and then move on to the elements that appear more than once.

Example: Lets use the reaction of sodium metal reacting with water to produce aqueous sodium hydroxide and hydrogen gas. Give the balanced chemical equation for this reaction.

Solution:

1.) Sodium metal should be represented as $Na(s)$ and should appear on the left-hand side of the arrow.
2.) Water is a liquid under normal conditions and should be represented as $H_2O(l)$.
3.) The sodium hydroxide is the combination of a metal and a polyatomic ion. You must recall that the hydroxide ion has a negative one charge and will thus combine with sodium in a ratio of 1 to 1. The sodium hydroxide on the products side would be represented as $NaOH(aq)$.
4.) The last product is hydrogen gas. Hydrogen forms one bond (HONC) and will therefore form a diatomic gas. This can be depicted as $H_2(g)$.

5.) The chemical equation for this reaction should be:

$Na(s) + H_2O(l) \rightarrow NaOH(aq) + H_2(g)$

6. The law of conservation of matter indicates that what goes in must come out. A quick look at the equation shows that two hydrogen atoms go into the reaction and three come back out again. The numbers of reactants and products must be manipulated until there are the same number of atoms going into the reaction as there are coming out of the reaction.
7. The balanced chemical equation will be:

$2\ Na(s) + 2\ H_2O(l) \rightarrow 2\ NaOH(aq) + H_2(g)$

e. The mole. In chemistry, the mole and Avogadro's number are used to represent 6.022×10^{23} particles. One mole of an element gives a direct correlation between the atomic mass of one atom of that element in amu and the molar mass of that element in grams. To determine how much a molecule weighs, it is simple enough to determine how many of each type of element is present. Once that is known, one mole of that molecule will be the combined mass in grams for each of the elements. The periodic table gives the mass in grams for one mole of any element. Where moles are invaluable is in the relationships expressed by chemical equations. The stoichiometry of a reaction gives the ratio of reactants to products. You did this in the previous section when you balanced the chemical equation. What you were really doing was determining the stoichiometry for the reaction. Using a combination of stoichiometry , the mole concept, and dimensional analysis, you can determine any combination of factors concerning a chemical reaction that you desire. A flow chart that emphasizes how you can accomplish the conversion is shown below.

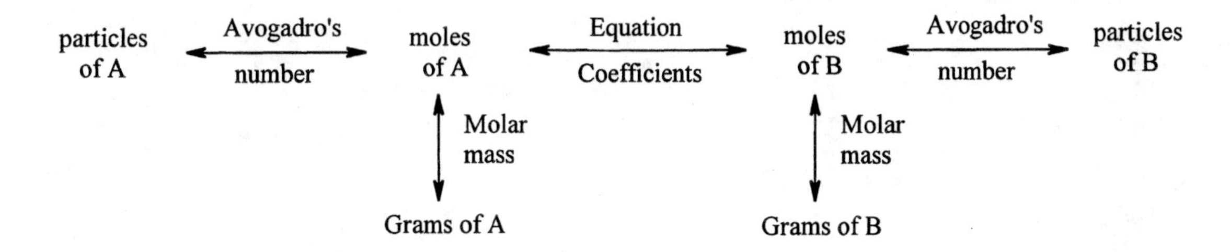

Solutions to Odd Numbered Problems:

1. A chemical formula shows the composition or structure of the smallest whole-number ratio of ions or atoms present in a compound. The symbol for each element in that compound is given and the abundance of each ion or element in the compound is shown by a numerical subscript.

3. The formula H_2O_2 indicates that there are two atoms of hydrogen bonding to two atoms of oxygen.

5. a. MgO contains one magnesium ion with a plus two charge and one oxyide ion with a minus two charge.
   b. Ag_2S contains two silver ions with plus one charges each and one sulfide ion with a negative two charge.
   c. BaI_2 contains one barium ion with a plus two charge and two iodide ions with negative one charges.
   d. CrF_3 contains one chromium ion with a plus three charge and three fluoride ions with negative one charges.

7. a. PCl_5 contains one atom of phosphorus and five atoms of chlorine.
   b. CS_2 contains one atom of carbon and two atoms of sulfur.
   c. N_2H_4 contains two atoms of nitrogen and four atoms of hydrogen.
   d. H_3PO_4 contains three atoms of hydrogen, one atom of phosphorus, and four atoms of oxygen

9. These compounds will be ionic if they are composed of a metal and a nonmetal and they will be covalent if they are composed of only nonmetals.
   a. KI will be ionic, because potassium is a metal while iodine is a nonmetal.
   b. PCl_5 will be covalent because both phosphorus and chlorine are nonmetals.
   c. CBr_4 is covalent because both carbon and bromine are nonmetals.
   d. SrO is ionic because strontium is a metal and oxygen is a nonmetal.
   e. VF_3 is ionic because vanadium is a metal and fluorine is a nonmetal.

11. a. aluminum
    b. bromide
    c. nitride
    d. cesium
    e. ferric or iron(III)

13. a. sodium iodide
    b. magnesium oxide
    c. potassium nitride
    d. calcium bromide

15. a. Iron(II) oxide
    b. Copper(I) iodide

17. a. Cuprous sulfide
    b. Ferrous bromide
    c. Mercuric chloride

19. a. Cesium sulfide
    b. Beryllium fluoride
    c. Mercuric bromide or Mercury(II) bromide

21. a. KBr
    b. MgI_2
    c. Li_3P

23. a. Cu_2S
    b. HgF_2
    c. FeI_3

25. a. Calcium hydroxide
    b. Sodium sulfate
    c. Magnesium carbonate

27. a. $NaOH$
    b. SnF_2
    c. $Fe_3(PO_4)_2$

29. Covalent compounds are created when two nonmetals are joined by covalent bonds. Ionic compounds are created when a metal and a nonmetal are joined by ionic bonds. Covalent bonds involve a sharing of electrons where ionic bonds involve the giving and taking of electrons. Two common covalent compounds are water and carbon dioxide.

31. a. Diphosphorus tetraiodide
    b. Disulfur difluoride
    c. Boron trifluoride
    d. Chlorine monofluoride
    e. Diselenium dichloride

33. a. ICl
    b. BP
    c. SO_3
    d. SeS
    e. N_2O_3

35. a. SO
 b. ICl_3
 c. IF_5
 d. SeO_2
 e. S_2O_7

37. a. Burning wood is a chemical change.
 b. Melting ice is a physical change.
 c. Dissolving sugar into coffee is a physical change.
 d. The rusting of a nail is a chemical change.

39. A chemical equation is a concise symbolic representation of a chemical reaction.

41. Two molecules of mercury(II) oxide are the reactants.

43. There are two molecules of mercury(II) oxide, two atoms of mercury, and one molecule of oxygen in the equation. The coefficients are 2, 2, and 1.

45. The reactants are glucose ($C_6H_{12}O_6$) and oxygen (O_2).

47. The coefficients are one for glucose, six for oxygen, six for carbon dioxide, and six for water.

49. a. Reactants = one atom of sulfur and four atoms of oxygen; Products = one atom of sulfur and three atoms of oxygen. The reaction is not balanced.
 b. Reactants = two atoms of aluminum and six atoms of bromine; Products = two atoms of aluminum and six atoms of bromine. The reaction is balanced.
 c. Reactants = two atoms of carbon and four atoms of oxygen; Products = two atoms of carbon and four atoms of oxygen. The reaction is balanced.
 d. Reactants = one atom of aluminum, three atoms of nitrogen, ten atoms of oxygen, one atom of sodium, and one atom of hydrogen; Products = one atom of aluminum, six atoms of oxygen, three atoms of hydrogen, one atom of nitrogen, and one atom of sodium. The equation is not balanced.

51. a. A coefficient of 2 must be placed prior to the chlorine in the reactants to balance the equation.
 $Sn(s) + 2Cl_2(g) \rightarrow SnCl_4(l)$
 b. A coefficient of 2 must be placed prior to the hydrogen chloride in the products to balance the equation.
 $CH_4(g) + 2Cl_2(g) \rightarrow CH_2Cl_2(l) + 2HCl(g)$
 c. A coefficient of 2 must be placed prior to the potassium nitrate in the products to balance the equation.
 $Pb(NO_3)_2(aq) + 2KI(aq) \rightarrow 2KNO_3(aq) + PbI_2(s)$

53. a. $2Na(s) + Cl_2(g) \rightarrow 2NaCl(s)$
 b. $2K(s) + 2H_2O(l) \rightarrow 2KOH(aq) + H_2(g)$
 c. $2LiOH(s) + CO_2(g) \rightarrow Li_2CO_3(s) + H_2O(l)$

55. a. $2P(s) + 5Cl_2(g) \rightarrow 2PCl_2(s)$
 b. $Ba(OH)_2(aq) + 2HCl(aq) \rightarrow BaCl_2(aq) + 2H_2O(l)$
 c. $3Ca(NO_3)_2(aq) + 2K_3PO_4(aq) \rightarrow Ca_3(PO_4)_2(s) + 6KNO_3(aq)$

57. A mole is 6.022×10^{23} of anything. There will be 6.022×10^{23} atoms of zinc in one mole of zinc just as there will be 6.022×10^{23} apples in one mole of apples.

59. To solve this problem, multiple 1.68 mol Sn by 6.022×10^{23} atoms Sn per mol of Sn. The answer is 1.01×10^{24} atoms of Sn.

61. Divide the atoms of sulfur by Avogadro's number to determine the moles of S. The answer is 22.4 moles of S.

63.

$$0.478 \text{ mol } H_2O \times \frac{6.022 \times 10^{23} \text{ molecules of } H_2O}{1 \text{ mol } H_2O} = 2.88 \times 10^{23} \text{ molecules of } H_2O$$

$$2.88 \times 10^{23} \text{ molecules of } H_2O \times \frac{2 \text{ atoms H}}{1 \text{ molecule } H_2O} = 5.76 \times 10^{23} \text{ atoms of H}$$

$$2.88 \times 10^{23} \text{ molecules of } H_2O \times \frac{1 \text{ atom O}}{1 \text{ molecule } H_2O} = 2.88 \times 10^{23} \text{ atoms of O}$$

65. 1.66×10^{-1} moles of glucose would contain 1.00×10^{23} molecules of glucose. There would be 9.96×10^{-1} moles of atoms of both carbon and oxygen and 1.99 moles of atoms of hydrogen.

67. a. Take the mass of Uranium from the periodic table (238 amu) and multiply by 2.35 to determine that 2.35 moles of uranium will have a mass of 559 grams.
b. 2.3×10^{-7} mol Ag $\times$ 107.87 g Ag/1 mol Ag = 2.5×10^{-5} g Ag
c. 2.3 mol $F_2 \times 38.00$ g F_2/1 mol F_2 = 87 g F_2

69. a. 1.000 kg K = 1000 g K $\times$ 1 mol K/39.10 g K = 25.58 mol K
b. 25 μg Au = 2.5×10^{-5} g Au $\times$ 1 mol Au/196.97 g Au = 1.3×10^{-7} mol Au
c. 25.0 g $O_2 \times 1$ mol O_2/32.00 g O_2 = 7.81×10^{-1} mol O_2

71. Molar mass is the atomic mass of an element expressed in grams per mole. A hydrogen atom has a mass of 1 amu where the molar mass of a mole of hydrogen is one gram.

73. a. 10.81 g of B/mol
b. 2×126.90 g of I = 253.80 g of I_2/mol
c. 3 N atoms = 3×14.01 g/mol = 42.03 g N; 12 H atoms = 12×1.01 g H/mol = 12.12 g H; 1 P atom = $1 \times$ 30.97 g P/mol = 30.97 g P; 4 O atoms = 4×16.00 g O/mol = 64.00 g O; To determine the molar mass of $(NH_4)_3PO_4$, add the masses of each constituent. 42.03 g N + 12.12 g H + 30.97 g P + 64.00 g O = 149.12 g/mol $(NH_4)_3PO_4$.
d. 1.01 g H/mol + 12.01 g C/mol + 14.01 g N/mol = 27.03 g/mol HCN

75. a. 32 g HCN $\times$ 1 mol HCN/27.03 g HCN = 1.2 mol HCN
b. 41.3 g $Cu(NO_3)_2 \times 1$ mol $Cu(NO_3)_2$/187.57 g $Cu(NO_3)_2$ = 2.20×10^{-1} mol $Cu(NO_3)_2$
c. 23.5 mg KBr = 2.35×10^{-2} g KBr $\times$ 1 mol KBr/119 g KBr = 1.97×10^{-4} mol KBr

77. a. 0.83 mol $C_{12}H_{22}O_{11} \times 342.34$ g $C_{12}H_{22}O_{11}$/1 mol $C_{12}H_{22}O_{11}$ = 2.8×10^2 g $C_{12}H_{22}O_{11}$
b. 11.3 μmol PbS = 1.13×10^{-5}mol PbS $\times$ 239.27 g PbS/1 mol PbS = 2.70×10^{-3} g PbS
c. 350 mol $H_2 \times 2.02$ g H_2/1 mol H_2 = 7.1×10^2 g H_2

79. To solve this problem, you must use the stoichiometry or coefficients given in the balanced chemical equation. This problem is asking for a mole to mole conversion. You are converting from moles of NH_3 to moles of N_2 and then to moles of H_2. The coefficients give you the ratio. For every two moles of NH_3 that are formed, you will need to use three moles of H_2 and one mole of N_2.
3.0 mol $NH_3 \times 1$ mol N_2/2 mol NH_3 = 1.5 mol N_2 gas will be necessary to make 3.0 mol of NH_3.
3.0 mol $NH_3 \times 3$ mol H_2/2 mol NH_3 = 4.5 mol H_2 gas will be necessary to make 3.0 mol of NH_3.

81. To determine the number of grams of ammonia are produced from 10.0 g of hydrogen, you must first convert the grams of hydrogen to moles of hydrogen and then use the equation coefficients to determine the moles of

ammonia. Once you have moles of ammonia, the molar mass of ammonia will give you a direct conversion to grams of ammonia.

grams of A ⟷ moles of A ⟷ moles of B ⟷ grams of B

$$10.0 \text{ g } H_2 \times \frac{1 \text{ mol } H_2}{2.02 \text{ g } H_2} \times \frac{2 \text{ mol } NH_3}{3 \text{ mol } H_2} \times \frac{17.04 \text{ g } NH_3}{1 \text{ mol } NH_3} = 56.2 \text{ g } NH_3$$

83. This problem is asking that you convert from moles of A (sodium bicarbonate) to gram of B (sodium carbonate). The solution involves converting from moles of sodium bicarbonate to moles of sodium carbonate and then to grams of sodium carbonate.

    moles of A ⟷ moles of B ⟷ grams of B

$$10 \text{ mol } NaHCO_3 \times \frac{1 \text{ mol } Na_2CO_3}{2 \text{ mol } NaHCO_3} \times \frac{105.99 \text{ g } Na_2CO_3}{1 \text{ mol } Na_2CO_3} = 5 \times 10^2 \text{ g } Na_2CO_3$$

The answer to the question is that 10 mol of sodium bicarbonate would be sufficient to product 400.0 g of sodium carbonate.

85. a. Ammonium fluoride
    b. Ferrous sulfide or iron(II) sulfide
    c. Silicon monocarbide
    d. Aluminum hydroxide
    e. Phosphorus pentabromide
    f. Beryllium bromide

87. a. PbO
    b. P_2O_4
    c. $BaCO_3$
    d. HgS
    e. Si_3N_4
    f. $NaNO_3$

89. a. Lithium chloride
    b. $2Li(s) + 2HCl(aq) \rightarrow 2LiCl(aq) + H_2(g)$
    c. 125 g Li × (1 mol Li/6.94 g Li) × (2 mol LiCl/2 mol Li) × (42.39 g LiCl/1 mol LiCl) = 764 g LiCl

91. 175 g $Mg(OH)_2$ × (1 mol $Mg(OH)_2$/58.32 g $Mg(OH)_2$) × (1 mol MgO/1 mol $Mg(OH)_2$) × (40.30 g MgO/1 mol MgO) = 121 g MgO will be necessary to make 175 g of $Mg(OH)_2$.

Sample Test:

1. What is the difference between a chemical formula and a molecular formula?
   a. A chemical formula gives the number of all atoms in the compound.
   b. A molecular formula gives the number of all atoms in the compound.
   c. A chemical formula gives the smallest whole-number ratio of ions present in a compound.
   d. More than one correct response.

2. In the compound calcium chloride, what is the ratio of calcium to chlorine, respectively?
   a. 1 : 1   b. 1 : 2   c. 2 : 1   d. 2 : 2

3. How many atoms are present in a molecule with the molecular formula of $C_6H_{12}O_6$?
   a. 6   b. 12   c. 18   d. 24

4. In the compound $ZnCl_2$, what is the charge of the zinc ion?
   a. +1 b. +2 c. –1 d. –2

5. What information can you get from the structural formula of a molecule?
   a. The number of atoms present in the molecule.
   b. The types of elements present in the molecule.
   c. The connectivity of atoms in the molecule.
   d. All of the above.

6. What is the name for the compound $MgBr_2$?
   a. maganous bromide b. mangonic bromide c. magnesium bromide d. magnesium(II) bromide

7. What is the charge on tin in the compound $Sn(SO_4)_2$?
   a. +2 b. +4 c. -2 d. -4

8. Give the formula for Vanadium(V) oxide.
   a. VO
   b. V_2O
   c. V_5O_2
   d. V_2O_5

9. Give the formula for Aluminum chloride
   a. AlCl b. $AlCl_2$ c. $AlCl_3$ d. Al_3Cl

10. Give the formula for ammonium bromide.
   a. ABr
   b. NH_4Br
   c. NH_4Br_4
   d. $Br(NH_4)$

11. What will be the coefficients needed to balance the following chemical equation?
   $$\_\_\_NH_3 + \_\_\_O_2 \rightarrow \_\_\_N_2 + \_\_\_H_2O$$
   a. 1, 1, 1, 1 b. 2, 1, 1, 3 c. 4, 3, 1, 6 d. 4, 3, 2, 6

12. In a chemical equation, what is on the left side of the reaction arrow and what is on the right side, respectively?
   a. products, reactants
   b. reactants, products
   c. solids, liquids
   d. solids, gases

13. What will be the coefficients needed to balance the following chemical equation?
   $$\_\_\_FeI_2 + \_\_\_Cl_2 \rightarrow \_\_\_FeCl_3 + \_\_\_I_2$$
   a. 1, 1, 1, 1 b. 1, 3, 1, 1 c. 2, 3, 2, 3 d. 2, 3, 2, 2

14. What is the chemical equation for the reaction of sodium with water to yield sodium hydroxide and hydrogen?
   a. $Na + H_2O \rightarrow NaOH + H_2$
   b. $2Na + H_2O \rightarrow 2NaOH + H_2$
   c. $2Na + 2H_2O \rightarrow 2NaOH + H_2$
   d. $2Na + 2H_2O \rightarrow 2NaOH + 2H_2$

15. For the reaction in problem 14, assign the state of matter present for the reactants and products, respectively.
   a. Solid, aqueous, liquid, gas
   b. Liquid, aqueous, aqueous, gas
   c. Solid, liquid, aqueous, gas
   d. Solid, liquid, liquid, liquid

16. What is the molar mass of caffeine ($C_8H_{10}N_4O_2$)?

17. How many oxygen atoms are there in 1.25 moles of $CaCO_3$?

18. When iron metal is mixed with oxygen gas, solid Ferric oxide is produced. Give the balanced chemical equation for the reaction.

19. The combustion reaction for ethanol (C_2H_6O) is given below. If 3.20 mol of ethanol is burned in the presence of excess oxygen, how many molecules of carbon dioxide will be produced?

    $C_2H_6O(l) + 3O_2(g) \rightarrow 2\ CO_2(g) + 3H_2O(l)$

20. How many grams of oxygen are present in 1.23 moles of ammonium sulfate?

<u>Answers to Sample Test:</u>

1. d
2. b
3. d
4. b
5. d
6. c
7. b
8. d
9. c
10. b
11. d
12. b
13. d
14. c
15. c
16. 194.22 g/mol
17. 2.26×10^{24} atoms of oxygen
18. $4Fe(s) + 3O_2(g) \rightarrow 2Fe_2O_3(s)$
19. 3.85×10^{24} molecules of CO_2
20. 78.7 g O

Chapter 5: Gases, Liquids, and Solids

Chapter Objectives:

1. To understand that matter can exist as a solid, liquid, or gas depending on the amount of energy exerted upon it.
2. To understand the impact of energy gains and loses on the temperature of substances.
3. To learn the nature of gases.
4. To determine the volume, pressure, temperature, or moles of a gas using the gas laws.
5. To differentiate between the difference in the intermolecular forces that hold together solids, liquids, and gases.

Key Terms: The key terms are listed in the order in which they are encountered in the Chapter.

States of matter. Matter exists as solids, liquids, and gases.

Changes in state. A physical change that results in the inter-conversion of matter between a solid, liquid, or gas.

Melting. A physical change of state that results in the conversion of a solid to a liquid.

Melting point. The temperature at which a solid is converted to a liquid.

Freezing. A physical change of state that results in the conversion of a liquid to a solid.

Freezing point. The temperature at which a liquid is converted to a solid.

Vaporization. The physical change of state that results in the conversion of a liquid to a gas.

Normal boiling point. The temperature at which a liquid is converted to a gas at one atmosphere of pressure.

Condensation. The physical change of state that results in the conversion of a gas to a liquid.

Sublimation. The physical change of state that results in the conversion of a solid to a gas.

Deposition. The physical change of state that results in the conversion of a gas directly to a solid.

Heat of fusion (ΔH_{fusion}). The amount of heat, in cal/g, a substance gains when it melts or the amount of heat a substance loses when it freezes.

Heat of vaporization. The amount of heat gained or lost when a liquid vaporizes or a gas condenses, respectively.

Specific heat. The amount of heat needed to raise the temperature of 1 g of a substance by 1°C.

Gas. Matter that has no definite shape and no definite volume.

Pressure. The force applied to a given area expressed in units of pounds per square inch (psi), inches of mercury (in Hg), millimeters of mercury (mm Hg), atmospheres (atm), torr, or pascals (Pa).

Standard temperature and pressure (STP). The values that are considered to be standard or common are a temperature of 0°C (273 K) and a pressure of 1 atm (760 torr).

Boyle's law. The relationship $P_1V_1=P_2V_2$ that indicates the product of pressure and volume for a gas remains constant.

Charle's Law. The relationship $V_1/T_1=V_2/T_2$ that indicates the volume of a gas at a certain temperature in Kelvin will go up or down as the temperature goes up or down.

Guy-Lussac's Law. The relationship $P_1/T_1=P_2/T_2$ that indicates the pressure of a gas at a certain temperature in Kelvin will go up or down as the temperature goes up or down.

Combined Gas Law. The relationship $(P_1V_1)/T_1=(P_2V_2)/T_2$ is a combination of Boyle's law and Charle's law.

Avogadro's Law. The relationship $V_1/n_1=V_2/n_2$ that indicates the volume of a gas will increase with more particles or moles of gas.

Ideal Gas Law. The relationship PV=nRT that is the composite of the Combined Gas law and Avogadro's law.

Molar volume. The volume (22.4 L) occupied by 1 mol of any gas at Standard Temperature and Pressure (273 K, 1 atm).

Universal Gas Constant (R). The value 0.0821 L · atm/K · mol that is used to solve the Ideal Gas law.

Dalton's law of partial pressures. The total pressure of a mixture of gases is equal to the sum of the pressures individually exerted by all the gases in the mixture ($P_{total}=P_1 + P_2 + \ldots P_n$).

Intramolecular bond. A bond that is present within a molecule.

Intermolecular bond. A bond that is present between different molecules; also referred to as intermolecular forces.

Dipole-dipole interaction. A weak attractive force between the partially positive end of one polar molecule and the partially negative end of another polar molecule.

Hydrogen bond. A special kind of dipole-dipole interaction between a very electronegative atom (O, N, and F) in one molecule and a hydrogen atom that is covalently bound to a very electronegative atom in another molecule.

London Dispersion Forces. A force of attraction in nonpolar molecules that involves the interaction between a temporary dipole and an induced dipole.

Solid. Matter that has definite shape and a definite volume.

Ionic solid. The force holding the particles of an ionic solid together is the interaction between anions and cations that constitutes ionic bonding.

Molecular solids. A solid made up of molecules that are held together by dipole-dipole interactions, hydrogen bonds, or London dispersion forces.

Network solids. A solid held together by covalent bonds between each atom.

Metallic solids. A solid that is held together by a sharing of valence electrons between atoms that gives the metal the ability to conduct electricity, be ductile, and be malleable.

Liquids. Matter that has a definite volume but an indefinite shape.

Chapter 5 Objective Details:

a. **Changes in State of Matter.** Matter exists as solids, liquids, or gases. When heat is added or taken away from matter, a change of state will occur if the heat added or taken away is sufficient. To determine whether or not the heat is sufficient, the heat of fusion and heat of vaporization can be determined for any substance. The heat of fusion is the amount of heat in calories per gram that a substance gains when it melts, or the amount of heat a substance loses when it freezes. The expression for heat of fusion is $q = m \times \Delta H_{fusion}$, where q is heat in calories, m is the mass of the substance in grams, and ΔH_{fusion} is the heat of fusion in calories per gram of substance. The heat of vaporization for a substance has the same formula, except the term ΔH_{fusion} is replaced by ΔH_{vap}. The units remain the same and the problem solutions are most easily handled by dimensional analysis. An important factor to remember is that changes in state are physical changes. If you melt a solid, you can return the solid to its original state by freezing it again.

b. Changes in Temperature. The key to this section is the formula for heat gain or loss during a temperature change. The heat gain or loss is equal to the product of the mass of the substance times the specific heat of the substance times the change in temperature of the substance represented by the equation: $q = m \times SH \times \Delta T$. The units for q are calories, m is in grams, SH is in calories per gram degree Celsius, and ΔT is in degrees Celsius. This last term, ΔT, is significant because when a change in state occurs, as discussed in the previous section, the temperature remains constant. One way to rationalize this concept is to think of a glass of ice water. The temperature of the water will remain 0°C until all of the ice is melted.

c. Gases. This section introduced the idea that a gas is a substance with no definite shape or volume. A gas will expand to fill a container regardless of its size. The reason gases do this is because they exert a pressure on their surroundings. Pressure is measured in many different units. The problem is that all of the different units are still used and frequently encountered. The idea of standard temperature and pressure (STP) is to give a common set of values for temperature and pressure. The standard temperature can be expressed as 0°C, 32°F, or 273 K. The most common of these is 273 K. The standard pressure can be expressed as 1 atm, 14.7 psi, 760 torr, 760 mm Hg, or 29.92 in Hg. The pressure unit you will encounter the most in chemistry will be atmospheres.

d. Gas Laws. There are seven gas laws to remember. This may seem like a lot to keep straight, but if you start with Boyle's law that relates pressure and volume and add that to Charle's law that relates volume and temperature, you will find that you are at the Combined Gas law. The Combined Gas law doesn't take into account the number of moles of gas in the mixture, so if you add Avogadro's law, you will find that you have just derived the Ideal Gas law. The things that you will need to keep straight with the Ideal Gas law is that 1 mol of gas at STP (273 K, 1 atm) will occupy a volume of 22.4 L, and the Universal Gas Constant is 0.0821 L · atm/K · mol. Another note is that when you deal with an equation involving temperature, you must convert the temperature to Kelvin ($T_K = T_C + 273$). One way to ensure success in any problem solving situation is to always include the units in the calculation. If you don't remember an equation, the units can help you figure out whether or not you have the formula correct. Make sure the units cancel and your answer is in the correct units. A chart of the gas laws is given below to show the similarities between the different equations.

Gas Law	Equation
Boyle's Law	$P_1V_1=P_2V_2$
Charle's Law	$V_1/T_1=V_2/T_2$ (T must be in Kelvin)
Gay-Lussac's Law	$P_1/T_1=P_2/T_2$ (T must be in Kelvin)
Combined Gas Law	$(P_1V_1)/T_1=(P_2V_2)/T_2$ (T must be in Kelvin)
Avogadro's Law	$V_1/n_1=V_2/n_2$ (n = moles)
Ideal Gas Law	PV=nRT (P in atm, V in L, R = 0.0821 L · atm/K · mol, T in K)
Dalton's Law of Partial Pressures	$P_{total}=P_1 + P_2 + \ldots P_n$

e. Solids and Liquids. Solids are defined as matter that has definite shape and a definite volume. The forces that hold solids together both intramolecular and intermolecular. Intramolecular forces are the ionic or covalent bonds that are present within a molecule. Intermolecular forces hold the molecules together. Intermolecular forces between molecules can be dipole-dipole interactions, hydrogen bonds, or London dispersion forces. There is a fine line between the definition of dipole-dipole interactions and hydrogen bonds. The distinction between the two is that dipole-dipole interactions involve the attraction between oppositely "partial" charges on different molecules; hydrogen bonds involve only the extremely electronegative elements oxygen, nitrogen, or fluorine. The common solids are ionic, molecular, network, and metallic. Liquids differ from solids in that their particles are not packed as closely together. There is more kinetic energy associated with the particles of liquids than there is with the particles of solids.

Solutions to Odd Numbered Problems:

1. Changes in state are physical changes. If you melt a solid, you can return the solid to its original state by freezing it again.

3. The normal boiling point for a substance is the temperature at which a liquid is converted to a gas at one atmosphere of pressure. The change of state where a gas is converted to a liquid is called condensation.

5. a. The substance is water and it is evaporating (vaporization).
   b. The layer may be fat that is going from a liquid to a solid upon cooling (freezing).
   c. The dew is water that is going from the gaseous form to the liquid form by condensation.

7. Heat of fusion.

9. The question is asking for the amount of energy in calories to melt a mass of 75 g of ice at a temperature of 0°C. This is a plug-in problem where you are asked to solve for q in the formula: $q = m \times \Delta H_{\text{fusion}}$. The value of ΔH_{fusion} comes from Table 5.3. $q = (75\ \text{g}) \times (79.7\ \text{cal/g})$ The units of grams cancel and the answer is 6.0×10^3 cal.

11. The heat of vaporization.

13. The question is asking for the amount of energy in calories to vaporize 455 g of water at a temperature of 100°C. This is a plug-in problem where you are asked to solve for q in the formula: $q = m \times \Delta H_{\text{vap}}$. The value of ΔH_{vap} comes from Table 5.4. $q = (455\ \text{g}) \times (540\ \text{cal/g})$ The units of grams cancel and the answer is 2.46×10^5 cal.

15. a. $q = m \times \Delta H_{\text{vap}} = (11.5\ \text{g})(159\ \text{cal/g}) = 1.83 \times 10^3$ cal.
    b. $q = m \times \Delta H_{\text{fusion}} = (37.5\ \text{g})(79.7\ \text{cal/g}) = 2.99 \times 10^3$ cal
    c. $q = m \times \Delta H_{\text{vap}} = (425\ \text{g})(138\ \text{cal/g}) = 5.87 \times 10^4$ cal
    d. 183 mg = 0.183 g
       $q = m \times \Delta H_{\text{fusion}} = (0.183\ \text{g})(2.8\ \text{cal/g}) = 0.512$ cal

17. The specific heat for any substance is the amount of heat needed to raise the temperature of 1 g of the substance by 1°C.

19. Using the equation $q = m \times SH \times \Delta T$, we can fill in the mass of the substance at $m = 3.00 \times 10^3$ g, the specific heat for water from Table 5.5 is given as $SH = 1.00$ cal/g °C, and the change in temperature is given as $\Delta T =$ 39°C. Solving for q gives the answer as 1.17×10^5 calories of energy will be absorbed by the water.

21. a. $q = m \times SH \times \Delta T$; $q = (24\ \text{g})(1.00\ \text{cal/g °C})(46\text{°C}) = 1.1 \times 10^3$ cal
    b. $q = m \times SH \times \Delta T$; $q = (225\ \text{g})(0.500\ \text{cal/g °C})(43\text{°C}) = 4.84 \times 10^3$ cal
    c. $q = m \times SH \times \Delta T$; $q = (385\ \text{g})(0.694\ \text{cal/g °C})(58\text{°C}) = 1.55 \times 10^4$ cal

23. a. $q = 5.04$ cal
    b. $q = 6.20 \times 10^3$ cal
    c. This problem must be done in two steps because of the change of state that occurs when solid water turns to liquid water. We must first solve for the energy associated with the temperature change. This is the same process as above. $q = m \times SH \times \Delta T$; $q = (875\ \text{g})(0.500\ \text{cal/g °C})(14\text{°C}) = 6.13 \times 10^3$ cal. The second step to the problem is to determine the energy associated with the change in state. For this, the heat of fusion for water must be known. The formula to use is $q = m \times \Delta H_{\text{fusion}}$. The heat of fusion for water can be found in Table 5.3 as 79.7 cal/g. The energy associated with the change in state is $(875\ \text{g})(79.7\ \text{cal/g}) = 6.97 \times 10^4$ cal. Combining the energy associated with the temperature change to the energy associated with the change of state, we can determine the total energy of the process to be 7.58×10^4 cal.

25.

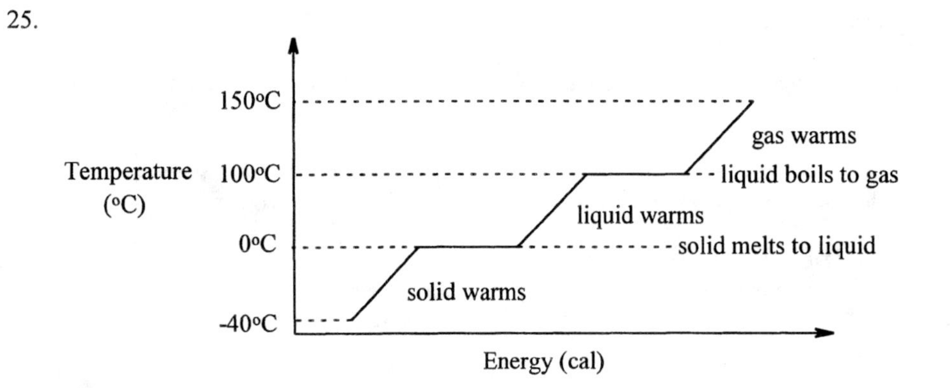

27. Matter that has no definite shape or volume is in the gaseous state. Three examples of substances or mixtures that are in the gaseous state include oxygen, methane, and helium.

29. Gas particles are in constant motion. This motion results in constant collisions between gas particles and the walls of the container that surround the gas. These frequent collisions exert pressure. If you double the number of particles in a container, you will double the number of collisions. This will result in a doubling of pressure.

31. STP stands for Standard Temperature and Pressure. The standard temperature can be expressed as 0°C, 32°F, or 273 K. The standard pressure can be expressed as 1 atm, 14.7 psi, 760 torr, 760 mm Hg, or 29.92 in Hg.

33. a. 746 Torr
    b. 0.982 atm
    c. 29.4 in Hg

35. Boyle's law shows the relationship between the volume of a gas and the pressure exerted upon or by the gas. Boyle's law states that when no other factors change, the pressure of a gas times the volume of the gas equals a constant. The symbols P and V are used to represent pressure and volume, respectively. The numbers 1 and 2 are used as subscripts to indicate the initial and final conditions, respectively. Boyle's law is dependent on the temperature of the gas remaining constant. If the temperature were to change, Boyle's law would no longer apply because volume is also dependent on temperature.

37. The pressure will be three times its original value. The volume will double if the pressure is halved.

39. a. $P_2 = (P_1V_1)/V_2 = 611$ torr
    b. $V_2 = (P_1V_1)/P_2 = 805$ mL
    c. $P_2 = (P_1V_1)/V_2 = 4.6$ atm
    d. $V_2 = (P_1V_1)/P_2 = 0.79$ L

41. $V_2 = (P_1V_1)/P_2 = 10.9$ L

43. Charle's law shows how the temperature and volume of a gas are related; $(V/T) =$ constant. The law is only valid if pressure remains constant because the volume of a gas is also dependent on the pressure of the surroundings.

45. When heat is applied to a gas, the particles move faster and the gas occupies a greater volume. If heated, a balloon will expand and the gas would occupy a greater volume.

47. a. $T_2 = (V_2T_1)/V_1 = (2611\ \text{L} \times 291\ \text{K})/2575\ \text{L} = 295\ \text{K} = 22°\text{C}$
    b. $V_2 = (V_1T_2)/T_1 = (83.26\ \text{mL} \times 525\ \text{K})/467\ \text{K} = 93.6\ \text{mL}$
    c. $V_2 = (V_1T_2)/T_1 = (81.7\ \text{L} \times 365.6\ \text{K})/476\ \text{K} = 62.8\ \text{L}$
    d. $T_2 = (V_2T_1)/V_1 = (277\ \text{mL} \times 311\ \text{K})/316\ \text{mL} = 273\ \text{K} = 0°\text{C}$

49. $V_2 = (V_1T_2)/T_1 = (27\ L \times 304\ K)/287\ K = 29\ L$

51. The volume of the balloon will increase. A higher temperature will give the gas particles in the balloon more kinetic energy and a reduced pressure will allow for less resistance. Both factors will contribute to the balloon increasing in volume.

53. The Combined Gas law should be used to solve this problem.
$V_2 = (P_1V_1T_2)/(T_1P_2) = (1.6\ atm \times 7.2\ L \times 238\ K)/(283\ K \times 2.2\ atm) = 4.4\ L$

55. The Combined Gas law should be used to solve this problem.
$P_2 = (P_1V_1T_2)/(T_1V_2) = (755\ torr \times 61.9\ mL \times 301\ K)/(294\ K \times 52.7\ mL) = 908\ torr$

57. The Combined Gas law should be used to solve this problem.
$V_2 = (P_1V_1T_2)/(T_1P_2) = (753\ torr \times 432\ L \times 265\ K)/(300\ K \times 516\ torr) = 557\ L$

59. Avogadro's law takes into account the number of particles of gas present; (V/n) = constant. Charle's and Boyle's laws do not include moles of gas.

61. Avogadro's law should be used to solve this problem. $V_2 = (V_1n_2)/n_1 = (51\ L \times 3.3\ mol)/2.6\ mol = 65\ L$

63. The Ideal Gas law gives the relationship PV=nRT that is the composite of the Combined Gas law and Avogadro's law.

65. Solve the Ideal Gas law for pressure and then substitute in the appropriate values given in the problem. Don't forget to convert the temperature to kelvin.
$P = (nRT)/V = (0.63\ mol \times 0.0821\ L \cdot atm/K \cdot mol \times 295\ K)/10.0\ L = 1.5\ atm$

67. Solve the Ideal Gas law for moles and then substitute in the appropriate values given in the problem. The units must be liters, atmospheres, and kelvin. 1 L = 1000 mL; 1 atm = 760 torr; $T_K = T_C + 273$
$n = (PV)/(RT) = (1.02\ atm \times 0.345\ L)/(0.0821\ L \cdot atm/K \cdot mol \times 291\ K) = 1.47 \times 10^{-2}\ mol$

69. Air is composed of nitrogen, oxygen, and several other gases. Atmospheric pressure is the sum of the pressures that all of the individual gases exert on their surroundings. The contribution of any one of the gases is known as the partial pressure for that gas.

71. The partial pressure for oxygen is 11 atm and the partial pressure for nitrogen is 16 atm.

73. The forces that occur between molecules in solids and liquids are intermolecular forces.

75. A hydrogen bond is a special kind of dipole-dipole interaction between a very electronegative atom (O, N, and F) in one molecule and a hydrogen atom that is covalently bound to a very electronegative atom in another molecule. Water and ice are common substances that have hydrogen bonding between their molecules.

77. Solid

79. Molecular solids may be held together by dipole-dipole interaction, hydrogen bonds, or London dispersion forces.

81. Metals are good conductors of electricity because they share their valence electrons with other metal atoms. This sharing of electrons results in a sea of electrons that can easily transfer an electrical current.

83. Liquid

85. a. The first step is to convert ice at -35°C to ice at 0°C.

$q = m \times SH \times \Delta T$; $q = (25.0\ g)(0.500\ cal/g\ °C)(35°C) = 438\ cal$

b. The second step is to convert ice from 0°C to water at 0°C.

$q = m \times \Delta H_{fusion} = 25.0\ g \times 79.7\ cal/g = 1.99 \times 10^3\ cal$

c. The third step of the problem is to convert water from 0°C to water at 100°C.

$q = m \times SH \times \Delta T$; $q = (25.0\ g)(1.00\ cal/g\ °C)(100°C) = 2.50 \times 10^3\ cal$

d. The fourth step is to convert water to steam.

$q = m \times \Delta H_{vap} = 25.0\ g \times 540\ cal/g = 1.35 \times 10^4\ cal$

e. The fifth step is to convert steam at 100°C to steam at 145°C.

$q = m \times SH \times \Delta T$; $q = (25.0\ g)(0.48\ cal/g\ °C)(45°C) = 540\ cal$

f. The final step is to total the energy required for each process.

$438\ cal + 1.99 \times 10^3\ cal + 2.50 \times 10^3\ cal + 1.35 \times 10^4\ cal + 540\ cal = 1.90 \times 10^4\ cal$

87. This problem requires the Ideal Gas law. Remember to convert to kelvin and atmospheres.

$n = (PV)/(RT) = (0.991\ atm \times 45.0\ L)/(0.0821\ L \cdot atm/K \cdot mol \times 297\ K) = 1.83$ mol of O_2

Once you have moles of oxygen, you will need to use Avogadro's number to get the number of oxygen molecules and then multiply that by two to get the number of oxygen atoms.

1.83 mol $O_2 \times (6.02 \times 10^{23}$ molecules O_2/mol $O_2) = 1.10 \times 10^{24}$ molecules of O_2 and 2.20×10^{24} atoms of oxygen

Sample Test:

1. After swimming, you exit the water and immediately feel cold. Why?
   a. The water on your skin is condensing and that requires heat.
   b. The water on your skin is vaporizing and that gives off heat.
   c. The water on your skin is subliming and that gives off heat.
   d. The water on your skin is vaporizing and that requires heat.

2. Frost-free freezers were developed because large deposits of ice were common in earlier generations. The ice build-up was due to what change in state?
   a. evaporation b. condensation c. sublimation d. deposition

3. A common way to sacrifice a good Corning Revere Ware pot is to steam vegetables. If you put 300 mL of water into a pot to steam broccoli, and you boil-off all of the water, how much energy was generated? The density of water is 1.00 g/mL, the heat of fusion of water is 79.7 cal/g, and the heat of vaporization of water is 540 cal/g.
   a. 1.62×10^5 cal b. 4.30×10^4 cal c. 2.39×10^4 cal d. 1.29×10^7 cal

4. Which change will require the greatest amount of energy?
   a. Freezing water
   b. Freezing a bottle of whiskey
   c. Boiling water
   d. Condensing methane

5. The specific heat of ice is 0.500 cal/g °C and that of water is 1.00 cal/g °C. Will it take more energy to heat 5.00 g of ice from -57°C to -12°C or to heat 2.75 mL of water from 23°C to 66°C? How much more energy?
   a. ice; 5.75 cal b. water; 5.75 cal c. ice 165.9 cal d. water; 165.9 cal

6. If 500 mL of water is heated from 23°C to 110°C, a change of state is associated with the temperature change. How much energy will be required for this temperature change to occur? $\Delta H_{vap} = 540$ cal/g; $SH_{water} = 1.00$ cal/g °C; $SH_{steam} = 0.48$ cal/g °C; density of water = 1.00 g/mL
   a. 3.11×10^5 cal b. 1.62×10^5 cal c. 4.35×10^4 cal d. 2.09×10^4 cal

7. A typical mountain bike tire should be inflated to 40.0 psi. Express this value in torr.
   a. 588 torr b. 3.04×10^4 torr c. 2.07×10^3 torr d. 760 torr

8. If your lungs hold 3.54 L of air at sea level (1.00 atm), how much air will they hold when you are scuba diving twenty meters under the surface of the ocean at a pressure of 4.35 atm?
   a. 15.4 L b. 0.814 L c. 1.23 L d. 0.177 L

9. If you inflate a balloon, for a friend's birthday party, to a volume of 1.50 L at a temperature of 75°F, what is the volume of the balloon in the party room that is at a temperature of 100°F?
   a. 1.57 L b. 2.00 L c. 1.13 L d. 1.43 L

10. The pressure inside of a car tire is 35 psi at a temperature of 25°C. What will be the pressure inside of the tire if the car is moving at 75 miles per hour and the friction of the road has heated the tire to 43°C?
    a. 33 psi b. 20 psi c. 37 psi d. 60.2 psi

11. A weather balloon is filled with helium gas to 498 L at 1.02 atm and a temperature of 22.6°C. If the balloon is released, and it shrinks to 182 L at 2.50 atm, how cold is it at this higher elevation?
    a. 17.2°F b. 68.4°F c. -380°F d. 38.1°F

12. An oxygen cylinder is used to inflate the bag on a nonrebreather mask. The bag is filled to 1.75 L with 0.538 moles of O_2. If a patient takes a breath of O_2 from the bag and the bag decreases in volume by 0.56 L, how many moles of gas has the patient inhaled?
    a. 0.366 mol O_2 b. 0.172 mol O_2 c. 0.253 mol O_2 d. 0.538 mol O_2

13. What will be the pressure exerted by 0.142 moles of nitrogen gas in a 500 mL container at a temperature of 40°C?
    a. 9.33×10^{-4} atm b. 7.30×10^{-3} atm c. 0.933 atm d. 7.30 atm

14. A little boy chewing gum blows a bubble. The air from his lungs is blown into the bubble at a pressure of 1.23 atm. Of the 1.23 atm, there are 0.76 atm of nitrogen and 0.15 atm of oxygen. What is the combined pressure of the other gases in the bubble?
    a. 1.23 atm b. 0.91 atm c. 0.32 atm d. 0.23 atm

15. A 50 cc syringe filled with oxygen gas is depressed to 39.5 cc by exerting a pressure of 985 torr. What was the initial pressure of the gas in the syringe?
    a. 0.75 atm b. 1.03 atm c. 1.64 atm d. 1.89 atm

16. Nitrogen exists as a gas under normal conditions. If nitrogen is taken to very low temperatures it will turn into a liquid. What forces would hold liquid nitrogen together?
    a. dipole-dipole interactions b. hydrogen bonds c. London dispersive forces d. ionic bonding

17. Which of the following molecules would be held together by dipole-dipole interactions with other of the same molecules?
    a. HF b. NaCl c. HNO_3 d. H_2

18. How are solids and liquids different?
    a. Solids have only dipole-dipole interactions.
    b. Liquids have only hydrogen bonding interactions.
    c. The intramolecular forces are stronger in solids than in liquids.
    d. The intermolecular forces are stronger in solids than in liquids.

19. How much energy will be required to raise the temperature of 125 mL of ethyl alcohol from 54°F to 75°F if the specific heat of ethyl alcohol is 0.581 cal/g °C and the density is 0.79 g/mL?
    a. 671 cal b. 850 cal c. 1100 cal d. 1200 cal

20. How many atoms of hydrogen will there be in 374 mL of gaseous sulfuric acid (H_2SO_4) at 94.3°F and 873 mm of Hg?
    a. 5.65×10^{-26} H atoms

b. 1.02×10^{22} H atoms
c. 2.05×10^{22} H atoms
d. 1.82×10^{26} H atoms

Answers to Sample Test:

1. d
2. d
3. a
4. c
5. b
6. a
7. c
8. b
9. a
10. c
11. a
12. b
13. d
14. c
15. b
16. c
17. c
18. d
19. a
20. c

Chapter 6: Solutions, Dispersions, and Suspensions

Chapter Objectives:

1. To learn the factors that affect the solubility of solutes in solutions.
2. To learn the units of concentration and how to use them.
3. To understand the differences between solutions, dispersions, and suspensions.
4. To become familiar with common colligative properties.

Key Terms: The key terms are listed in the order in which they are encountered in the Chapter.

Solution. A homogeneous mixture of two or more pure substances.

Homogeneous mixture. A mixture that is the same throughout.

Heterogeneous mixture. A mixture that has a variable composition.

Solvent. The most abundant component of a solution.

Aqueous Solution. A solution prepared with water as the solvent.

Solute. The less abundant component or components of a solution.

Solubility. The tendency of a substance to dissolve into another substance.

Ion-dipole interactions. The interaction that occurs between a polar molecule and an ion that leads to the solubility of ionic compounds in polar solvents.

Like dissolves like. This statement indicates that polar and ionic substances dissolve in polar solvents while nonpolar substances dissolve in nonpolar solvents.

Saturated solution. A solution that contains the maximum amount of dissolved solute.

Unsaturated solution. A solution that can still dissolve more solute.

Concentration. The amount of solute present for a given quantity of solution.

Molarity (M). The number of moles of solute per liter of solution.

Mass/mass (m/m) percent. The mass/mass percent of a solution is the mass of solute in grams divided by the mass of solution in grams multiplied by 100.

Mass/volume (m/v) percent. The mass/volume percent of a solution is the mass of solute in grams divided by the volume of the solution in milliliters multiplied by 100.

Volume/volume (v/v) percent. The volume/volume percent of a solution is the volume of solute in milliliters divided by the volume of the solution in milliliters multiplied by 100.

Dilution. The addition of solvent to lower the concentration of solute in a solution.

Henry's law. A gas in contact with a liquid will have a constant concentration in that liquid which is directly proportional to the pressure of the gas.

Nitrogen narcosis. A state of altered consciousness that results from higher than normal amounts of nitrogen dissolved in the blood.

The bends. The severe pain and discomfort caused when nitrogen boils out of blood as pressure is rapidly decreased.

Suspension. An unstable mixture of larger particles suspended in a liquid.

Filtration. The process by which large particles are removed from a suspension.

Centrifugation. A method of filtration that uses centrifugal force to separate large particles from a suspension.

Dispersions (Colloids). A homogeneous mixture of two or more substances where the particles of the substances are too big to form a true solution but small enough to be unaffected by gravity.

Tyndall effect. The effect observed when light is deflected as it passes through a dispersion.

Colligative properties. The properties of a solution that are dependent only on the amount of solute in the solution, not on the nature of the solute.

Boiling-point elevation. A colligative property that involves the increase in boiling-point for a solution that accompanies the addition of more solute molecules.

Freezing-point depression. A colligative property that involves the decrease in freezing-point for a solution that accompanies the addition of more solute molecules.

Semipermeable membrane. A barrier that will only allow the passage of some substances.

Osmosis. The movement of water molecules through a semipermeable membrane into a solution.

Osmotic pressure. The pressure that must be applied to a solution to prevent the movement of water across a semipermeable membrane.

Osmolarity. A measure of concentration that considers all solute particles outside a solution.

Isotonic. The condition that exists when the concentration of solute is the same inside a cell as it is inside of a cell.

Hypotonic. The condition that exists when the concentration of solute is lower outside of the cell than it is inside of the cell.

Hypertonic. The condition that exists when the concentration of solute is higher outside of the cell than it is inside of the cell.

Hemodialysis. A treatment for kidney failure that removes wastes and restores salt and water balance.

Crenate. Cell shrinkage caused by the cell being placed in a hypertonic solution.

Hemolysis. The rupturing of a cell associated with the cell swelling in a hypotonic solution.

Dialysis. A separation technique that uses semipermeable membranes

Chapter 6 Objective Details:

a. Solutions and Solubility. A solution is homogeneous mixture of two or more pure substances. The major component to a mixture is the solvent and the minor component(s) is the solute. Different solutes will vary in their solubility in a given solvent. The reason for the variation in solubility is due to intermolecular forces that contribute to a solute being dissolved in a solvent. It is a general rule that like dissolves like. Polar or ionic compounds will dissolve in polar solvents and nonpolar solutes will dissolve in nonpolar solvents. The universal solvent is water. Solutions that have water as the solvent are known as aqueous solutions.

b. Concentrations. We learned earlier that like dissolves like, but we didn't find out how to express how much dissolves. This section covered the aspect of expressing the amount of solute that will dissolve in a given amount of solution. If a solution contains enough of a solute such that no more of the solute will dissolve in that solvent, the solution is said to be saturated. An unsaturated solution will still be able to dissolve more solute. The most common way to express the amount of solute dissolved in a solution is the molarity of the solution. The molarity of a solution is the number of moles of solute per liter of solution. The point that needs to be made here is that the liters of solution takes into account both the solute and solvent. This point is especially important when solution concentrations are expressed as mass/mass percents, mass/volume percents, and volume/volume percents. If the concentration of a solution is known, a dilution is easily accomplished to reduce the amount of solute in the solution to a desired level. Henry's law gives the relationship between the concentration of a gaseous solute and pressure. The key equations for this section are given below.

Concentration Units	Equations
Molarity	1 M = 1 mole of solute/1 liter of solution = 1 mol/1 L
Mass/mass percent	% (mass/mass) = grams of solute/grams of solution × 100
Mass/volume percent	% (mass/volume) = grams of solute/milliliters of solution × 100
Volume/volume percent	% (volume/volume) = milliliters of solute/milliliters of solution × 100
Dilution	$C_1V_1 = C_2V_2$ (C = concentration; V = volume; 1 = initial; 2 = final)
Henry's law	Concentration of gas in liquid = constant × pressure of the gas

c. Dispersions and Suspensions. A solution can be thought of as a homogeneous mixture of two or more substances that can only be separated by distillation. As the solute particles get bigger, the homogeneous mixture is called a dispersion or colloid. Dispersions can be separated by filtration or centrifugation and will deflect a beam of light that is passed through it. A suspension involves solute particles that are large enough to be separated by gravity. Common methods to separate suspensions are filtration and centrifugation.

d. Colligative properties. Colligative properties are those that depend on the number of solute particles in a solution. Freezing-point depression is the result of increasing numbers of solute particles causing a solution to freeze at a lower temperature. Boiling-point elevation is the result of increasing numbers of solute particles causing a solution to boil at a higher temperature. Osmotic pressure is the pressure that must be applied to a solution to prevent the movement of water across a semipermeable membrane. The more solute present, the higher the pressure that must be applied to prevent water movement. The terms isotonic, hypotonic, and hypertonic are used to indicate when the concentration of solute is the same outside of a cell as inside, when solute concentrations are lower outside of a cell than outside, and when solute concentrations are higher outside of a cell than inside, respectively.

Solutions to Odd Numbered Problems:

1. A solution is a homogeneous mixture of two or more pure substances. Salt water is an example of a solution.

3. A solute is the less abundant component or components of a solution. Sodium chloride is the solute in a solution of salt water.

5. A tincture of iodine is a alcohol solution of a nonvolatile medicine. Alcohol is the solvent, so iodine must be the solute.

7. Dipole-dipole interactions or hydrogen bonds can occur between polar solvent molecules and polar solute molecules.

9. Water is a polar solvent. Each hydrogen atom has a partial positive charge and each oxygen atom has a partial negative charge. A sodium cation will be surrounded by the partial negative charge of the oxygen atoms in molecules of water. The result of this interaction is called an ion-dipole interaction.

11. a. Gasoline will be soluble in benzene but not water.
    b. Glucose will be soluble in water but not benzene.
    c. Potassium iodide will be soluble in water but not benzene.
    d. Methane will be soluble in benzene but not water.

13. Unsaturated

15. Saturated

17. The molarity of a solution is the number of moles of solute that are dissolved per liter of solution.
    1 M = 1 mol/1 L

19. a. (1.4 mol of NaCl)/(2.50 L of solution) = 0.56 M
    b. (0.025 mol I_2)/(0.125 L solution) = 0.200 M
    c. (0.185 mol HCl)/(4.5 L solution) = 0.041 M

21. (0.014 mol caffeine)/(0.225 L solution) = 0.062 M

23. The moles of solute can be determined by multplying the molarity of the solution by the volume of the solution in liters.
    a. (0.15 M NaCl) (2.0 L) = 0.30 mol NaCl
    b. (2.5 M HNO_3) × (0.275 L) = 0.69 mol HNO_3
    c. (2.75 M KCl) × (0.0416 L) = 0.114 mol KCl
    d. (1.1 M $(NH_4)_2SO_4$) × (450 L) = 5.0×10^2 mol $(NH_4)_2SO_4$

25. (0.10 M $AgNO_3$) × (0.100 L solution) = 0.010 mol $AgNO_3$

27. The first thing to do is to find out how many moles of each salt are needed to produce the desired concentration at a volume of 0.250 L. This can be done by multiplying the molarity of the solute by the volume of the solution in liters.
    For KBr: (0.10 M KBr) × (0.250 L) = 0.025 mol of KBr
    For NaCl: (0.25 M NaCl) × (0.250 L) = 0.063 mol NaCl
    An additional step to the problem would be to determine the number of grams of each solute are required. Both KBr and NaCl are ionic solids and it is therefore instructive to determine the number of grams of solid that will need to be dissolved. This is accomplished by multiplying the moles of each solute by the molecular weight of that solute to get to grams.
    For KBr: 0.025 mol KBr × 119 g KBr/1 mol KBr = 3.0 g KBr
    For NaCl: 0.063 mol NaCl × 58.44 g NaCl/1 mol NaCl = 3.7 g NaCl

29. a. [(2.6 g $AgNO_3$)/(125 g solution)] × 100 = 2.1% (m/m)
    b. [(114 g sucrose)/(1114 g solution)] × 100 = 10.2% (m/m)
    c. [(28.5 g Sn)/(95.0 g solder)] × 100 = 30.0% (m/m)

31. [1.5% (m/m) × 35 g ointment]/100 = 0.53 g zinc oxide

33. a. [(1.00 g ascorbic acid)/(150.0 mL)] × 100 = 0.667% (m/v)
    b. [(0.50 g creatin)/(200.0 mL)] × 100 = 0.25% (m/v)
    c. [(3.00 g NH_3)/(235 mL)] × 100 = 1.28% (m/v)

35. a. [(35 mL acetone)/(150 mL solution)] × 100 = 23% (v/v)
    b. [(18.5 mL alcohol)/(125 mL solution)] × 100 = 14.8% (v/v)

37. [(volume of solution) × (% v/v)] ÷ 100 = volume of solute
[(125 mL solution) × (22.5% v/v)] ÷ 100 = 28.1 mL alcohol

39. $C_1V_1 = C_2V_2$
[(1.0 M HCl)(2.5 L)] ÷ 12 M HCl = 0.208 L of concentrated HCl = 2.1×10^2 mL of concentrated HCl

41. [(2.0 M HNO_3)(0.085 L)] ÷ 6.0 M HNO_3 = 0.028 L of 6 M HNO_3 = 28 mL of 6 M HNO_3

43. Henry's law states that a gas in contact with a liquid will have a constant concentration in that liquid which is directly proportional to the pressure of the gas. By increasing the pressure of oxygen the patient breathes, the patients blood will absorb more oxygen.

45. A solution will not deflect a beam of light passing through it while a dispersion will deflect a beam of light. An example of a dispersion is blood plasma.

47. A suspension has larger solute particles than a dispersion. With time, the particles in a suspension will separate from the solvent while the particles in a dispersion will remain dispersed.

49. Gasoline is a solution because it is made of several different hydrocarbons mixed together.

51. The mixture could be either a solution or dispersion. Shining a beam of light through the mixture and looking for the Tyndall effect is one way to differentiate between a solution and a dispersion.

53. If you didn't shake a solid medication suspended in an aqueous solution, you would not receive an adequate dose of the medication. The solid medication in a suspension will settle to the bottom of the bottle with prolonged exposure to gravity.

55. The Tyndall effect is the effect observed when light is deflected as it passes through a dispersion. A sunbeam passing through a room gets deflected by dust particles in the air.

57. Colligative properties are properties of a solution that are dependent only on the amount of solute in the solution, not on the nature of the solute. Freezing-point depression is a colligative property. As the concentration of solute increases in a solution, the freezing-point of the solution will decrease.

59. The freezing point of a solution will decrease with the addition of a solute. As more solute is added, the freezing point will decrease further.

61. Osmotic pressure is the pressure that must be applied to a solution to keep water from moving across a semipermeable membrane. As the concentration of solute increases, the osmotic pressure also increases.

63. When an animal cell is placed in a hypotonic solution, it will swell until it ruptures. When an animal cell is placed in a hypertonic solution, it will shrink. When an animal cell is placed in an isotonic solution, it will neither swell nor shrink.

65. Colligative properties are properties of a solution that are dependent only on the amount of solute in the solution, not on the nature of the solute. The boiling-point elevation, freezing-point depression, and osmotic pressure would be affected equally by a 0.50 M aqueous solution of glucose as it would a 0.50 M aqueous solution of sucrose.

67. The contents of each can will get cold, but only the diet soda would have a chance of actually freezing. If the cans contained pure water, they would be expected to freeze at a degree or two below freezing, but soda and beer are solutions and will therefore have a lower freezing point that pure water.

69. A solution that is isotonic to blood will have the same total concentration of solutes as the blood itself.

71. 38.5 g $CaCl_2$ × (1 mol $CaCl_2$/110.98 g $CaCl_2$) = 0.347 mol $CaCl_2$
    a. 1 M = moles solute/L solution = 0.347 mol $CaCl_2$/1.00 L = 0.347 M $CaCl_2$
    b. 0.347 M $CaCl_2$ × (1 mol Ca/1 mol $CaCl_2$) = 0.347 M Ca^{2+} ion
    c. 0.347 M $CaCl_2$ × (2 mol Cl/1 mol $CaCl_2$) = 0.694 M Cl^- ion

73. 0.100 M sucrose × 1.00 L = 0.100 mol sucrose
    0.100 mol sucrose × (342 g sucrose/1 mol sucrose) = 34.2 g sucrose

75. Filtration using filter paper with spaces of 0.1 mm to 10 μm will not be sufficient to filter the particles from a dispersion. A dispersion has particles on the order of nanometers to submicrometer in size. These particles will pass through the filter.

Sample Test:

1. A solution of rubbing alcohol is 70% isopropanol and 25% water and 5% ethanol. What is the solute?
   a. Isopropanol b. Water c. Ethanol d. Both water and ethanol.

2. Identify the heterogeneous mixture.
   a. Salt and water b. Gas and oil c. Bacon grease and water d. Sugar and water

3. What is the main force that contributes to the solubility of NaCl in water?
   a. Hydrogen bonding
   b. Dipole-dipole interactions
   c. Ion-dipole interactions
   d. London-dispersive forces

4. Which change will require the greatest amount of energy?
   a. Freezing water
   b. Freezing a bottle of whiskey
   c. Boiling water
   d. Condensing methane

5. A 150 lb adult contains about 6.6 liters of blood in their body. Nitrogen makes up 78% of the air we breathe (4.46×10^{-2} mol N_2/1 L air). If a 150 lb adult inhales 125 mL of air and it is assumed that all of the nitrogen inhaled is absorbed by the blood, what will be the molarity of the nitrogen in this person's blood?
   a. 4.46×10^{-2} M b. 6.67×10^{-3} M c. 8.45×10^{-4} M d. 3.68×10^{-2} M

6. The battery in your car contains 1.2 L of 12 M H_2SO_4. How many molecules of H_2SO_4 are present in the battery?
   a. 14.4×10^{23} molecules b. 8.67×10^{24} molecules c. 6.02×10^{24} molecules d. 6.02×10^{22} molecules

7. Salt water from the ocean is 7% (m/v) sodium chloride. How many grams of sodium chloride are there in 1.57 L of ocean water?
   a. 1.10×10^{-1} g NaCl b. 110 g NaCl c. 1.10×10^{3} g NaCl d. 1.10×10^{4} g NaCl

8. In the movie The Patriot, Mel Gibson melted-down metal toy soldiers to use as ammunition for his musket. If the composition of a toy soldier is 24.5 g Pb, 15.4 g Sn, and 34.5 g Fe, what is the mass/mass percent of lead (Pb) in a toy soldier?
   a. 20.7% (m/m) b. 32.9% (m/m) c. 46.4% (m/m) d. not enough information

9. Ethanol is commonly consumed as a recreational beverage. The amount of ethanol is typically expressed in terms of a volume/volume percentage. A 375 mL bottle of Crown Royal blended Canadian whiskey is 40% ethanol by volume. How many milliliters of pure ethanol are in the 375 mL bottle of blended whiskey?
   a. 0.150 mL ethanol b. 9.38 mL ethanol c. 150 mL ethanol d. 225 mL ethanol

10. A pot of coffee is brewed first thing in the morning and contains 2.00 L of coffee that is 0.357 M caffeine. A fire alarm goes off and lasts for 3 hours. Upon returning to the coffeepot, the volume has been reduced to 1.34 L due to the evaporation of water. What is the molarity of caffeine in the coffeepot assuming the only change that occurred was the loss of water?
    a. 0.957 M caffeine b. 0.714 M caffeine c. 0.533 M caffeine d. 0.239 M caffeine

11. Hyperventilation involves rapid deep breaths of air that decrease carbon dioxide levels and increase oxygen levels in the blood. The effect of this is that the alkalinity of the blood increases due to the decreased partial pressure of carbon dioxide. The old method of treatment for hyperventilation was to have the patient breath in and out of a paper bag. Why would this method work in theory?
    a. Breathing into a paper bag would allow the patient to better utilize the oxygen in the air.
    b. The partial pressure of oxygen would increase in the bag.
    c. The partial pressure of carbon dioxide would increase in the bag.
    d. It is an old method so it must not have worked.

12. Pepto Bismo contains the active ingredient bismuth subsalicylate. The directions indicate that pepto bismo should be shaken well prior to use. Pepto bismo is a murky pink mixture. Which of the following terms would best describe pepto bismo?
    a. Solution b. Colloid c. Dispersion d. Suspension

13. Epsom salt is the common name given to magnesium sulfate. Epsom salt is used as a laxative and soaking aid. Magnesium sulfate is dissolved in water and used as a soak or oral medication. Adults should dissolve two level teaspoons of the salt in an 8 ounce glass of water. The mixture appears clear and colorless. Based on your knowledge of chemistry, what type of mixture has been formed?
    a. Solution b. Colloid c. Dispersion d. Suspension

14. If someone puts sugar in the gas tank of your car, why is it bad?
    a. Sugar will have no effect.
    b. Sugar will dissolve in the gas and lower the freezing point.
    c. Sugar will dissolve in the gas and raise the boiling point.
    d. Sugar is polar and will not dissolve in gas because it is nonpolar.

15. Hypoglycemia occurs when blood sugar levels fall to abnormally low levels. The treatment is to give the patient glucose. What happens to the glucose as it enters the body?
    a. The glucose will create an isotonic solution in the body.
    b. The glucose will create a hypertonic solution in the body.
    c. The glucose will create a hypotonic solution in the body.
    d. The glucose will have no effect on the body.

16. What is the osmolarity of a 0.50 M $CaCl_2$ solution?
    a. 0.50 b. 1.0 c. 1.5 d. 3.0

17. What types of molecules could be separated by dialysis?
    a. Salts b. Urea c. Proteins d. More than one correct response

18. How many moles of oxygen are in 500 mL of a 0.472 M solution of potassium permanganate ($KMnO_4$)?
    a. 0.944 mol O b. 0.472 mol O c. 0.236 mol O d. 59.0 mol O

19. A man was recently arrested in Virginia for marketing homemade moonshine. The individual was distilling ethanol from corn and selling it to the Yankees in the north country. For every 750 mL of moonshine, there was found to be 575 mL of pure ethanol. What was the volume/volume percent of solute?
    a. 0.77% (v/v) solute b. 76.7% (v/v) solute c. 23.3% (v/v) solute d. 130% (v/v) solute

20. Compare the osmotic pressure of a 0.50 M NaCl solution to a 0.25 M $CaCl_2$ solution?
    a. The NaCl solution will have a higher osmotic pressure.
    b. The $CaCl_2$ solution will have a higher osmotic pressure.

c. The osmotic pressure will be the same for NaCl as it is for $CaCl_2$
d. No correct answer.

Answers to Sample Test:

1. d
2. c
3. c
4. b
5. c
6. b
7. b
8. b
9. c
10. c
11. c
12. d
13. a
14. d
15. b
16. c
17. d
18. a
19. c
20. a

Chapter 7: Chemical Reactions

Chapter Objectives:

1. To understand the differences between acids, bases, and salts.
2. To know when water is an acid or a base and how it ionizes.
3. To comprehend the significance of pH and how it relates to everyday solutions.
4. To determine the outcome of an acid-base reaction.
5. To understand the transfers in oxidation and reduction reactions.
6. To understand the factors involved in chemical reactions.

Key Terms: The key terms are listed in the order in which they are encountered in the chapter.

Salts. The ionic compounds that form when acids and bases react with one another.

Acid. A proton donor. Any compound that can give up a hydrogen ion.

Proton. A hydrogen atom can lose an electron to form a hydrogen ion which is referred to as a proton.

Ionization. The process that converts a molecule into ions.

Base. A proton acceptor. Any compound that can accept a hydrogen ion.

Dissociate. The process that separates an ionic compound into ions.

Hydronium ion (H_3O^+). The ion that forms when water acts as a base and accepts a proton.

Strong acids. Acid molecules that ionize 100% in water.

Weak acids. Acids molecules that ionize less than 100% in water.

Double arrow. A double arrow in a chemical equation means that the reaction is reversible or less than 100% of the reactants are used up in the reaction.

Single arrow. A single arrow in a chemical equation indicates that 100% of the reactants are used up in the course of the reaction.

Strong base. A hydroxide-containing compound that dissociates in water to yield a high concentration of hydroxide ion (OH^-).

Weak base. A compound that produces only small quantities of hydroxide ion in the presence of water.

Neutral solution. A solution that contains equal numbers of hydrogen ions and hydroxide ions.

Acidic solution. A solution that contains a larger amount of hydrogen ion than hydroxide ion.

Alkaline solution. A basic solution that contains a larger amount of hydroxide ion than hydrogen ion.

Ion-product constant of water (K_w). The number $1.0 \times 10^{-14} = [H^+][OH^-] = K_w$. This expression allows for the calculation of proton or hydroxide ion molar concentrations in water or aqueous solutions.

pH scale. A scale, with a range of 0 to 14, that is used to express hydrogen ion concentrations in aqueous solutions.

Neutralization. The reaction of a strong acid with a strong base to produce a salt and water.

Titration. Mixing an acid or base of unknown concentration with an acid or base of known concentration, in the presence of a pH indicator, to the point of neutralization.

pH indicator. A compound that has different colors at different pH values.

Equivalent. The amount of an acid that produces one mole of hydrogen ions or the amount of base that produces one mole of hydroxide ions.

Equivalent mass. The mass in grams of an acid or base that yields one mole of hydrogen ion or hydroxide ion, respectively.

Diprotic acid. An acid that can ionize to yield two hydrogen ions.

Dibasic base. A base that can dissociate to yield two moles of hydroxide ion.

Buffer. A solution that can resist changes in pH when an acid or base is added to it.

Oxidation-reduction (redox) reactions. A chemical reaction in which there is a loss and gain of electrons by the reactants.

Oxidation. The loss of electrons.

Reduction. The gain of electrons.

Precipitation reaction. A chemical reaction that produces an insoluble or solid product.

Precipitate. In a chemical reaction, this is solid that falls out of solution as the result of a precipitation reaction.

Exothermic reaction. A reaction that produces heat.

Endothermic reaction. A reaction that requires heat or that takes up heat.

Enthalpy of reaction. The amount of energy given up or taken in by a chemical reaction.

Heat of reaction. The same as the enthalpy of reaction.

Kinetics. The rate of a reaction or how fast a reaction occurs.

Energy of activation. The energy that must be provided by collisions before reactants can become products.

Catalyst. A substance that speeds up the rate of a reaction by lowering the activation energy, but is not used up in the reaction.

Chemical equilibrium. The state of a chemical reaction where the forward reaction occurs at exactly the same rate as the reverse reaction.

LeChatelier's principle. If a chemical equilibrium is disturbed, changes will occur that attempt to reestablish equilibrium.

Chapter 7 Objective Details:

a. **Acids, bases, and salts.** Salts are ionic compounds. If you remember the earlier discussions (Section 3.3) regarding metals and nonmetals coming together to make a salt, that's what we're looking at again. The additional bit of information is that the metal comes from a base and the nonmetal comes from an acid. To summarize, a salt is formed from the reaction of an acid with a base. An acid is a proton donor and a base is a proton acceptor. Through the course of your studies, you will be able to get away with remembering a few examples of acids and bases. Strong acids will ionize 100% in aqueous solution (water) while strong bases will dissociate 100% in aqueous solution. If the compound doesn't ionize or dissociate 100%, then it isn't strong. There are six strong acids and eight strong bases that you will encounter. If you know the fourteen strong acids and bases, you can assume anything else is a weak acid or weak base. Acids can be identified because the acid proton is always listed first in the formula. In discussions of weak bases, the only one that is commonly encountered is ammonia (NH_3). If a hydroxide (OH^-) is present in a formula, think strong base. The strong bases are easy to identify because the metal ion will be from either group 1A or group 2A with a hydroxide ion. The periodic table can be very helpful in the identification of acids and bases. The following table should also help.

Type of Substance	Name of Substance	Chemical Formula
Strong acid (ionize 100% in water)	Hydrochloric acid	HCl
	Hydrobromic acid	HBr
	Hydroiodic acid	HI
	Nitric acid	HNO_3
	Perchloric acid	$HClO_4$
	Sulfuric acid	H_2SO_4
Weak acid (ionize less than 100% in water)	Acetic acid	$HC_2H_3O_2$
	Carbonic acid	H_2CO_3
	Phosphoric acid	H_3PO_4
Strong base (dissociate 100% in water)	Lithium hydroxide	$Li(OH)$
	Sodium hydroxide	$Na(OH)$
	Potassium hydroxide	$K(OH)$
	Rubidium hydroxide	$Rb(OH)$
	Cesium hydroxide	$Cs(OH)$
	Calcium hydroxide	$Ca(OH)_2$
	Strontium hydroxide	$Sr(OH)_2$
	Barium hydroxide	$Ba(OH)_2$
Weak base (produces less than 100% hydroxide ion upon dissociation)	Ammonia	NH_3

b. **Ionization of water.** The ion-product constant of water, K_w, allows for the determination of hydrogen ion or hydroxide ion molar concentration. $K_w = 1.0 \times 10^{-14} = [H^+][OH^-]$. The brackets around the hydrogen ion and the hydroxide ion indicate the molar concentrations of each of the ions. There are three types of solutions that will be encountered. A neutral solution will have equal concentrations of hydrogen ions and hydroxide ions. The ion-product constant for water indicates that the concentration of hydrogen ion will be equal to the concentration of hydroxide ion and that the concentration of each ion will be 1.0×10^{-7} M. An acidic solution will have a greater concentration of hydrogen ion than hydroxide ion and a basic or alkaline solution will have a higher concentration of hydroxide ion than hydrogen ion.

c. **pH.** The pH of a solution is determined by the equation: $pH = -\log [H^+]$. The pH scale is far more convenient to use than the hydrogen ion concentration. The pH scale ranges from 0 to 14, in aqueous solution, with seven being in the middle. This indicates that a solution with a pH of 7 will be neutral. If the pH of the solution is higher than seven, the solution will be alkaline just as a solution with a pH of less than seven will be acidic. With this knowledge, you have a way to check to see if you are correct in solving problems. In the previous section, we used the ionization constant of water to determine whether or not a solution was acidic, alkaline, or neutral. In this section, we have a different method to get to the same information. You should always check to ensure that the answer you arrive at is consistent regardless of the way in which you determined it. The author eluded to significant figures and explained why a hydrogen ion concentration of 1.54×10^{-4} M would give a pH of 3.813 rather than 3.81 or 3.8125. The hydrogen ion concentration is 0.000154 M where there are three significant figures to the right of the decimal place. Your answer should also have three significant figures to the right of the decimal place. You can always go backwards from pH to hydrogen ion concentration by the use of the formula: $[H^+] = 10^{-pH}$. It is recommended that you always take the time to check your answer.

d. Acid-base reactions. A neutralization reaction involves mixing a strong acid and a strong base to produce a salt and water. A neutralization reaction in the presence of a pH indicator is known as a titration. When a solution of unknown acidity is mixed with a base of known alkalinity, the concentration of the unknown acid can be determined. The volume and molarity of the base are multiplied to get the moles of base needed for neutralization. If the acid is known, a balanced chemical equation can be written to determine the stoichiometry (Section 4.5) or ratio of acid to base. This ratio allows for the conversion of moles of base to moles of acid. The number of moles of acid divided by the liters of solution gives the molarity or concentration of the previously unknown acid.

Example: In a nitric acid (HNO_3)-potassium hydroxide (KOH) acid-base titration, 43.7 mL of 0.279 M KOH is required to neutralize 100.0 mL of HNO_3. What is the molarity of the acidic solution?

Solution:

1.) Determine the number of moles of base. You know that 43.7 mL of 0.279 M KOH was used in the titration. Multiply the volume in liters by the molarity to get the moles of KOH.

$0.0437 \text{ L KOH} \times 0.279 \text{ mol KOH/1 L KOH} = 0.0122 \text{ mol KOH}$

2.) Write the balanced chemical equation for the neutralization reaction and determine the ratio of acid to base. Remember that a strong acid and a strong base will always produce a salt and water.

$HNO_3(aq) + KOH(aq) \rightarrow KNO_3(aq) + H_2O(l)$

The balanced chemical equation indicates that 1 mol KOH will react with 1 mol HNO_3.

3.) Convert the moles of base to moles of acid.

0.0122 mol KOH were used in the titration of 0.0122 mol HNO_3

4.) Determine the molarity of the acid by dividing the moles of acid by the volume of the acidic solution.

$0.0122 \text{ mol } HNO_3/0.1000 \text{ L solution} = 0.122 \text{ M } HNO_3$

5.) The answer to the problem is the concentration of unknown acid is 0.122 molar.

In the above example, one mole of nitric acid produced one mole of hydrogen ion and one mole of potassium hydroxide produced one mole of hydroxide ion. It can be said that one equivalent of acid reacted with one equivalent of base. An equivalent is the amount of acid that produces one mole of hydrogen ion or the amount of base that produces one mole of hydroxide ion. An equivalent mass is the mass of an acid or base that yields one mole of hydrogen ion or hydroxide ion. This is an issue for diprotic and triprotic acids and for dibasic bases. Diprotic and triprotic acids are acids that can ionize to produce two and three moles of hydrogen ions, respectively. A dibasic base can dissociate to produce two moles of hydroxide ion. To determine the equivalent mass of an acid or base, divide the molar mass (in grams) of the acid or base by the number of equivalents of that acid or base.

Example: What mass of barium hydroxide ($Ba(OH)_2$) yields one mole of OH^-?

Solution:

1.) Identify the type of base. Is barium hydroxide mono- or dibasic?

$Ba(OH)_2 \rightarrow Ba^{2+} + 2\ OH^-$

The base is dibasic.

2.) What is the mass of one mole of the base?

$MW_{Ba} = 137.33 \text{ g/mol} + MW_O = 2(16.00 \text{ g/mol}) + MW_H = 2(1.01 \text{ g/mol}) = 171.35 \text{ g/mol}$

3.) Divide the molar mass of the base by the number of equivalents.

$171.35 \text{ g } Ba(OH)_2/2 \text{ equivalents} = 85.68 \text{ g} = \text{equivalent mass}$

Buffers are solutions that resist changes in pH. A buffer is the mixture of a weak acid and the salt of that weak acid, or a weak base and the salt of that weak base. It should be ingrained that strong acids and strong bases are not good buffers because they ionize or dissociate 100% in aqueous solution. A good buffer will contain an acid component to absorb the addition of base and a base component to absorb the addition of acid.

e. Oxidation-reduction and precipitation reactions. Redox reactions involve the loss or gain of electrons by the reactants. The origin of the term oxidation meant to gain oxygen. The definition of oxidation has been broadened over the years to mean a loss of electrons. Reduction is just the opposite of oxidation. The original definition was a loss of oxygen that has been broadened to mean a gain of electrons. A few ways to remember oxidation and reduction are oilrig and LEO says GER. Oilrig is used to represent oxidation is loss, reduction is gain. LEO says GER means loss of electrons, oxidation; gain of electrons, reduction. The method for remembering oxidation and reduction is up to you.

In general, metals become oxidized and nonmetals become reduced. From previous discussions in Chapter 3, we learned that metals will tend to lose electrons to become positively charged. Now we are calling that process oxidation. In a similar manner, nonmetals tend to gain electrons to become negatively charged. That process will now be referred to as reduction.

Precipitation reactions result in the formation of an insoluble product. Using the information learned in the previous chapter, you could say that a precipitation reaction yields a suspension.

f. **Properties of chemical reactions.** Reactions that give off energy are known as exothermic and those that require energy are known as endothermic. An easy way to tell if a reaction is endothermic or exothermic is to observe whether the reaction is warm or cold. A warm reaction or one that gives off heat is exothermic where a reaction that is cold or takes in heat is endothermic. The rate at which a chemical reaction occurs is known as the kinetics of the reaction. The kinetics of a reaction explains how particles interact. The rate of a chemical reaction is dependent on the number of collisions that occur between particles and the average energy of those collisions. For a chemical reaction to occur, an activation energy barrier must be overcome. Catalysts are compounds that lower the activation energy of a chemical reaction but remain unchanged themselves at the completion of the reaction. When a chemical reaction has the same forward rate of reaction as reverse rate, the reaction is said to be in equilibrium. A reaction in equilibrium will attempt to maintain that equilibrium. If more reactants are added to the reaction, more products will be produced until equilibrium is reestablished. The reverse is also true. If more products are added to a chemical reaction at equilibrium, the reverse reaction will form more reactants until the equilibrium is reestablished. The process of reestablishing equilibrium is known as LeChatelier's principle.

Solutions to Odd Numbered Problems:

1. An acid is a proton donor. Two examples of acids are acetic acid and sulfuric acid.

3. A salt is an ionic compound. Salts are produced from the reaction of an acid with a base. Table salt ($NaCl$), and baking soda ($NaHCO_3$) are two examples of salts in the home.

5. A chemical formula that begins with hydrogen identifies an acid.
   a. acid
   b. base
   c. salt
   d. acid
   e. acid
   Letters a, d, and e are acids.

7. A compound that causes a solution to feel slippery is typically an indication of a base. It may be assumed that sodium carbonate forms a base in solution.

9. Ionization

11. $HCl(aq) + H_2O(l) \rightarrow H_3O^+(aq) + Cl^-(aq)$

13. This textbook introduced hydrochloric acid (HCl), nitric acid (HNO_3), and sulfuric acid (H_2SO_4) as common strong acids.

15. This is a weak acid. If the acid were strong, 100% ionization would occur.

17. Formic acid is a weak acid, so a double arrow must be used in the chemical equation.

$$HCHO_2(aq) + H_2O(l) \rightleftharpoons CHO_2^-(aq) + H_3O^+(aq)$$

19. A weak base is a compound that produces only small quantities of hydroxide ion in the presence of water. Ammonia is an example of a weak base.

21. Pure water will contain a tiny number of both hydrogen ions and hydroxide ions.

23. Acidic

25. a. Basic
    b. Acidic
    c. Basic
    d. Neutral

27. K_w is the symbol for the ion-product constant of water. The numerical value of K_w at 25°C is 1.0×10^{-14}.

29. $[OH^-] = 1.0 \times 10^{-14}/[H^+]$
    a. $[OH^-] = 1.0 \times 10^{-14}/5.26 \times 10^{-12}\ M = 1.90 \times 10^{-3}\ M$
    b. $[OH^-] = 1.0 \times 10^{-14}/1.78 \times 10^{-9}\ M = 5.62 \times 10^{-6}\ M$
    c. $[OH^-] = 1.0 \times 10^{-14}/3.92 \times 10^{-5}\ M = 2.55 \times 10^{-10}\ M$
    d. $[OH^-] = 1.0 \times 10^{-14}/6.94 \times 10^{-4}\ M = 1.44 \times 10^{-11}\ M$

31. $[H^+] = 1.0 \times 10^{-14}/[OH^-]$
    a. $[H^+] = 1.0 \times 10^{-14}/9.04 \times 10^{-4}\ M = 1.11 \times 10^{-11}\ M$
    b. $[H^+] = 1.0 \times 10^{-14}/2.48 \times 10^{-9}\ M = 4.03 \times 10^{-6}\ M$
    c. $[H^+] = 1.0 \times 10^{-14}/1.32 \times 10^{-12}\ M = 7.58 \times 10^{-3}\ M$
    d. $[H^+] = 1.0 \times 10^{-14}/5.38 \times 10^{-6}\ M = 1.86 \times 10^{-9}\ M$

33. $pH = -\log [H^+]$
    a. $pH = -\log 4.04 \times 10^{-5}\ M = 4.394$
    b. $pH = -\log 7.418 \times 10^{-10}\ M = 9.1297$
    c. $pH = -\log 5.02 \times 10^{-2}\ M = 1.299$
    d. $pH = -\log 5.838 \times 10^{-9}\ M = 8.2337$

35. $[H^+] = 10^{-pH}$
    a. $[H^+] = 10^{-4.83} = 1.48 \times 10^{-5}\ M$
    b. $[H^+] = 10^{-6.07} = 8.51 \times 10^{-7}\ M$
    c. $[H^+] = 10^{-12.41} = 3.89 \times 10^{-13}\ M$
    d. $[H^+] = 10^{-8.273} = 5.333 \times 10^{-9}\ M$

37. a. $pH < 7.0$ (acid)
    b. $pH > 7.0$ (base)
    c. $pH < 7.0$ (acid)
    d. $pH > 7.0$ (base)

39. Most acidic → lemon juice, milk, blood, urine ← most basic

41.

pH	$[H^+]$	$[OH^-]$	Acid, Neutral, or Base?
6.41	3.9×10^{-7}	2.6×10^{-8}	acid
8.317	4.82×10^{-9}	2.07×10^{-6}	base
7.442	3.61×10^{-8}	2.77×10^{-7}	base

43. a. $HBr(aq) + NaOH(aq) \rightarrow NaBr(aq) + H_2O(l)$
    b. $Ba(OH)_2(aq) + 2\ HNO_3(aq) \rightarrow Ba(NO_3)_2(aq) + 2\ H_2O(l)$
    c. $H_3PO_4(aq) + 3\ KOH(aq) \rightarrow K_3PO_4(aq) + 3\ H_2O(l)$

45. $Mg(OH)_2(aq) + 2\ HCl(aq) \rightarrow MgCl_2(aq) + 2\ H_2O(l)$

47. A pH indicator is a compound that has different colors at different pH values. A pH indicator can be used in a titration to show when an acid has been neutralized. Upon neutralization, the pH of the solution will change in pH and when it does the pH indicator will change color.

49. a. 0.0417 L NaOH × 0.102 M NaOH = 0.00425 mol of NaOH
    b. $H_2SO_4(aq) + 2\ NaOH(aq) \rightarrow Na_2SO_4(aq) + 2\ H_2O(l)$
    c. 0.00425 mol NaOH × (1 mol H_2SO_4/2 mol NaOH) = 0.00213 mol H_2SO_4
    d. 0.00213 mol H_2SO_4/0.025 L solution = 0.0852 mol H_2SO_4/L
    e. Answer = 8.52×10^{-2} M H_2SO_4

51. a. 1
    b. 2
    c. 1
    d. 2

53. A buffer is a solution that can resist changes in pH when an acid or base is added to it. Buffered aspirin can remain neutral or unchanged in either an acidic or an alkaline environment.

55. a. Hydrofluoric acid is a weak acid and so the combination of a weak acid and the salt of that weak acid (KF) would be a good buffer.
    b. Nitric acid is a strong acid and would not be a good buffer.

57. An oxidation-reduction reaction is a chemical reaction in which there is a loss and gain of electrons by the reactants. An example is the reaction of solid magnesium in the presence of oxygen to yield magnesium oxide. The magnesium loses two electrons, which are gained by the oxygen.

59. Reduction is a gain of electrons. The metal gains a positive charge with each electron that it loses indicating oxidation. The non-metal gains a negative charge with each electron gained. Its oxidation number or charge decreases. This is reduction.

61. a. $2\ Ba(s) + O_2(g) \rightarrow 2\ BaO(s)$ where barium is oxidized and oxygen is reduced.
    b. $2\ Al(s) + 3\ Br_2(l) \rightarrow 2\ AlBr_3(l)$ where aluminum is oxidized and bromine is reduced.
    c. $Mg(s) + I_2(g) \rightarrow MgI_2(s)$ where magnesium is oxidized and iodine is reduced.

63. a. Carbon and hydrogen are oxidized; oxygen is reduced.
    b. Carbon and oxygen have been reduced and hydrogen has been oxidized.
    c. Carbon has been oxidized and zinc has been reduced.

65. As the water slowly evaporates, the concentration of salt in the solution will be greater than the water can handle. The salt will precipitate out of solution because its concentration has surpassed saturating conditions.

67. The enthalpy of a reaction is the amount of energy given up or taken in by a reaction.

69. a. The more often particles collide, the faster the rate of a reaction.
    b. The more energy particles contain when they do collide, the faster the rate of reaction.

71. A catalyst is a substance that speeds up the rate of a reaction without being used up in the reaction. Catalysts lower the activation energy of a reaction allowing the reaction to proceed more rapidly.

73. Chemical equilibrium is the state of a chemical reaction where the forward reaction occurs at exactly the same rate as the reverse reaction. Therefore it is true that the rate of the reverse reaction is the same as the rate of the forward reaction.

75. A buffer at equilibrium will have an acidic portion and an alkaline portion. The alkaline portion of the chemical expression will absorb the addition of acid. As the alkaline substance absorbs additional protons, it is converted to the acidic form. The absorption of acid will continue until the amount of acid and the amount of base are once again equivalent, at equilibrium, or the alkaline portion of the buffer is entirely used up.

77. Blood buffering is accomplished by the equilibrium between carbonic acid and bicarbonate. Carbon dioxide levels directly influence the formation of carbonic acid. The more carbon dioxide present in the blood as in the case of hypoventilation, the more carbonic acid that will be produced to reestablish the equilibrium. The opposite is also true. If there is a shortage of carbon dioxide in the blood, as in the case of hyperventilation, then levels of carbonic acid will drop and there will be an excess of the alkaline bicarbonate until the body can once again reestablish equilibrium.

Sample Test:

1. In the following reaction, what substance is the base?

   $HCHO_2(aq) + H_2O(l) \rightleftharpoons CHO_2^-(aq) + H_3O^+(aq)$

   a. $HCHO_2$ b. H_2O c. H_3O^+ d. There is no base in this reaction.

2. Identify the salt in the following equation.
   $NaOH(aq) + HCl(aq) \rightarrow NaCl(aq) + H_2O(l)$

   a. NaOH b. HCl c. NaCl d. H_2O

3. In question number one, a double arrow is used while in question number two a single arrow is used in the chemical equation. Why?
   a. The double arrow indicates complete ionization.
   b. The single arrow indicates complete reaction.
   c. The double arrow indicates complete dissociation.
   d. There is not difference between them.

4. A lake in the Adirondac mountains of upstate New York was found to have $[H^+] = 1.43 \times 10^{-5}$ M. What is the hydroxide ion concentration in the lake?
   a. 1.0×10^{-14} M b. 1.43×10^{-5} M c. 1.43×10^{9} M d. 6.99×10^{-10} M

5. What type of aqueous solution would have a hydroxide ion concentration of 6.54×10^{-9} M?
   a. Acidic b. Alkaline c. Neutral d. Not enough information.

6. Black coffee has a pH = 5.12. What is the hydroxide ion concentration of this coffee?
   a. 5.12 M b. 7.6×10^{-6} M c. 1.3×10^{-9} M d. 7.6×10^{8} M

7. If seawater from the ocean has a hydrogen ion concentration of 5.01×10^{-9}, what is the pH of seawater?
   a. 8.3 b. 5.7 c. 5.01 d. 7.0

8. A solution of ammonia is found to have a hydroxide ion concentration of 3.47×10^{-3} M. What is the pH of the solution?
   a. 1.154 b. 2.460 c. 11.00 d. 11.54

9. Many people will immediately throw baking soda on an acid spill, acid burn, or use a teaspoon full in a glass of water for acid indigestion. Baking soda is nothing more that sodium bicarbonate ($NaHCO_3$). When sodium bicarbonate is mixed with hydrochloric acid (HCl), what reaction takes place?

a. A strong acid and a strong base.
b. A strong acid and a weak base.
c. A weak acid and a strong base.
d. A weak acid and a weak base.

10. 25.0 mL of battery acid (sulfuric acid; H_2SO_4), removed from your car battery, was tested to determine the molarity of sulfuric acid. A pH indicator was observed to change color in the acidic solution after 57.8 mL of 0.250 M NaOH was added. What is the molarity of the acid from your car battery?
a. 2.89×10^{-4} M H_2SO_4 b. 0.250 M H_2SO_4 c. 0.289 M H_2SO_4 d. 0.578 M H_2SO_4

11. What mass of perchloric acid ($HClO_4$) is needed to provide 2.40 mol of hydrogen ion?
a. 2.40 g b. 41.86 g c. 101 g d. 241 g

12. Which of the following pairs of substances could function as a buffer system in aqueous solution?
1. HNO_3/$NaNO_3$ 2. NaCl/NaCN 3. NH_4Cl/NH_3 4. HF/NaF
a. 1,2,3,4 b. 1,2 c. 2,3 d. 3,4

13. When calcium is burned in the presence of oxygen, a bright flame is associated with the formation of calcium oxide. In the reaction: $2\ Ca(s) + O_2(g) \rightarrow 2\ CaO(s)$, what has happened to each calcium atom from reactants to products?
a. Calcium has been oxidized and lost two electrons.
b. Calcium has been oxidized and gained two electrons.
c. Calcium has been reduced and gained two electrons.
d. Calcium has been reduced and lost two electrons.

14. Most vehicles are made primarily of iron. When the iron in your vehicle combines with oxygen from the air, what causes it to rust?
a. Oxidation b. Reduction c. Precipitation d. Pollution

15. Sombra is a therapeutic pain relieving gel that contains capsaicin (cayenne pepper extract). When Sombra is applied to sore muscles, a burning sensation occurs. What type of reaction is occurring?
a. Oxidation b. Reduction c. Endothermic d. Exothermic

16. When baking soda and vinegar are mixed, the heat of reaction is –45 kcal/mol. What does this tell you about the reaction?
a. The reaction gives off heat and is endothermic.
b. The reaction gives off heat and is exothermic.
c. The reaction takes in heat and is endothermic
d. The reaction takes in heat and is exothermic.

17. It has been said that more collisions at a higher average energy will increase the rate of a reaction. If you were to mix 10 mL of ammonia with 12 mL of acetic acid, how could you speed up the rate of the reaction?
a. Heat the reaction.
b. Add 15 mL of ammonia rather than 10 mL.
c. Add 15 mL of acetic acid rather than 12 mL.
d. All of the above.

18. A catalyst will NOT...
a. Lower the activation energy for a reaction.
b. Change the equilibrium of a reaction.
c. Change the rate at which equilibrium is established.
d. Be used up in the chemical reaction.

19. The following chemical equation represents the phosphate buffer system in an aqueous environment:

$$H_2PO_4^- + OH^- \rightleftharpoons HPO_4^{2-} + H_3O^+$$

Which direction will the equilibrium shift to reestablish itself if additional hydroxide ions are added?
a. To the left. b. To the right. c. It will remain unchanged. d. Not enough information.

20. Hot tubs and swimming pools need to be pH balanced (buffered). Which of the following compounds would you add to a hot tub with a pH of 5.4 to get it in the desired range of 7.2-7.6?
a. HCl b. NaOH c. $NaHCO_3$ d. HNO_3

Answers to Sample Test:

1. b
2. c
3. b
4. d
5. a
6. c
7. a
8. d
9. b
10. c
11. d
12. d
13. a
14. a
15. d
16. b
17. d
18. b
19. b
20. c

Chapter 8: Saturated Hydrocarbons

Chapter Objectives:

1. To determine whether a compound is organic or inorganic.
2. To draw structures of alkanes and provide names for them.
3. To identify and name branched-chain alkanes.
4. To become familiar with the structure and nomenclature of cycloalkanes.
5. To understand the significance of the different shapes of organic molecules.
6. To learn the reactions and physical properties of alkanes.
7. To become familiar with functional groups common to organic chemistry.

Key Terms: The key terms are listed in the order in which they are encountered in the chapter.

Organic chemistry. The chemistry of carbon containing compounds.

Inorganic chemistry. The chemistry of all compounds that do not contain carbon.

Biochemistry. The chemistry of living systems.

Hydrocarbons. Organic compounds that contain only carbon and hydrogen atoms.

Alkanes. Hydrocarbons that contain only single bonds.

Saturated hydrocarbons. Hydrocarbons that are saturated with hydrogen; no more hydrogen can bind to carbon.

Straight-chain alkanes. An alkane having all of the carbon atoms linked by single bonds in one continuous chain.

Condensed structural formula. A way to represent the structure of molecules where some or all of the lines that represent bonds are not shown.

Molecular models. Images of organic molecules in three-dimensional space.

Ball-and-stick model. A ball is used to represent an atom and a stick is used to represent the bond between atoms in which bond angles and bond lengths are roughly to scale.

Space-filling model. A molecular image that provides the overall shape of a molecule but does not give a good representation of bond angles and bond lengths.

Perspective formula. A molecular representation that uses a solid wedge between atoms to indicate an atom coming out of the two-dimensional plane and a dashed wedge is drawn to an atom that is going into the two-dimensional plane of the paper.

Branched alkane. Any alkane that has one or more carbon atoms branching from the main chain of carbon atoms.

Isomers. Organic compounds that have the same composition but different structures and different chemical properties.

Substituent. A group (atom(s)) that is attached to the carbon atoms of the longest carbon chain in a molecule.

Alkyl group. A substituent that is formed by an alkane by removing one hydrogen atom.

Cycloalkanes. Saturated hydrocarbons that have a carbon skeleton in the form of a ring.

Positional isomers. Compounds that have the same molecular formula, but differ in the position of their substituents.

Geometric isomers. Molecules that differ in three-dimensional shape, but not in the order of attachment of their atoms or in the composition of their atoms.

cis-isomer. The prefix cis- is used to identify the geometric isomer that has both substituents on the same side of the ring.

trans-isomer. The prefix trans- is used to identify the geometric isomer that has both substituents on opposite sides of the ring.

Stereoisomers. Isomers that are different from one another only in the arrangement of atoms in three-dimensional space.

Chiral objects. Objects that are nonsuperimposable mirror images of one another.

Chiral compounds. Stereoisomers that are nonsuperimposable mirror images of one another.

Enantiomers. A pair of chiral molecules that are nonsuperimposable mirror images of one another.

Chiral center. An atom in a molecule that has four different substituents attached to it.

Configuration. The molecular shape of a molecule that is the result of the covalent bonding that is present between atoms in the molecule.

Conformation. The different arrangement of atoms in three-dimensional space achieved by rotation about single bonds.

Combustion. An oxidation-reduction reaction between a combustible material, like an alkane, and oxygen to produce carbon dioxide and water.

Ozone. The highly reactive form of oxygen with a molecular formula O_3 that is formed when molecular oxygen absorbs ultraviolet light.

Chlorofluorocarbons. Molecules that contain chlorine, fluorine, carbon, and sometimes hydrogen that are blamed for the disappearance of the ozone layer.

Halogenation. A light or heat catalyzed substitution reaction involving a halogen replacing a hydrogen atom on an alkane to produce a halogenated alkane and a hydrohalogen.

Alkyl halides. An alkyl group (alkane minus hydrogen) with a halogen attached.

Organohalogen. An organic molecule that contains halogen atoms.

Structural theory of chemistry. The properties of a compound are dependent upon the structure of the compound.

Functional groups. Atoms or groups of atoms that give a molecule its specific chemical and physical properties.

Heteroatom. A functional group or part of a functional group that are atoms other than carbon or hydrogen.

Chapter 8 Objective Details:

a. **Organic chemistry.** The chemistry of carbon containing compounds. Carbon containing compounds are present in all living organisms making organic chemistry one of the largest fields of chemistry. Carbon typically forms four covalent bonds and has an electronegativity of 2.5. These factors will be addressed in much greater detail.

b. Alkanes: the simplest organic compounds. Alkanes are saturated hydrocarbons that can be either straight-chained or branched. The straight-chain hydrocarbons have a nomenclature system that you will use throughout the remainder of the course. The prefixes, meth-, eth-, prop-, but-, pent-, hex-, hept-, oct-, non-, and dec- should be committed to memory as quickly as possible. These prefixes indicate the presence of one, two, three, four, five, six, seven, eight, nine, and ten carbon atoms, respectively. The suffix that indicates a saturated hydrocarbon is –ane. In organic chemistry, anytime a molecular name ends in –ane, you should think of a saturated hydrocarbon. One way to derive the formula of an alkane is to go by the formula, C_nH_{2n+2}, where n is equal to the number of carbon atoms. This formula indicates that if there are 3 carbon atoms in an alkane, there must be 8 hydrogen atoms. If you know the prefixes, then you can always derive the chemical formula. In this case, the formula for propane is C_3H_8. You try!

Example: What is the formula for hexane?

Solution:

1.) The prefix hex- indicates that there are six carbon atoms present.
2.) The suffix –ane indicates that the molecule is a saturated hydrocarbon.
3.) Using the formula C_nH_{2n+2}, where n=6, gives the formula for hexane as C_6H_{14}.

Once you feel confident with the nomenclature, it is then time to figure out how to represent these molecules. The options are structural formulas, condensed structural formulas, ball-and-stick models, space-filling models, and perspective formulas. You are already familiar with structural formulas from Section 4.1. Condensed structural formulas are based on the assumption that you know that carbon forms four bonds. If you know that, then you don't have to explicitly draw out every bond as you do in a structural formula. The term condensed comes from the structural formula being simplified. The ball-and-stick and space-filling models are much easier to visualize with models that you can hold in your hands. Perspective formulas are more common and you should become familiar with them. A comparison of the structural formula, condensed structural formula, and perspective formula for propane is given in Figure 1.

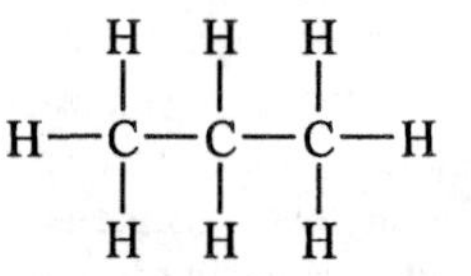

```
    H   H   H
    |   |   |
H — C — C — C — H
    |   |   |
    H   H   H
```

Structural formula

$CH_3CH_2CH_3$

$CH_3\text{-}CH_2\text{-}CH_3$

$H_3CCH_2CH_3$

Condensed structural formulas

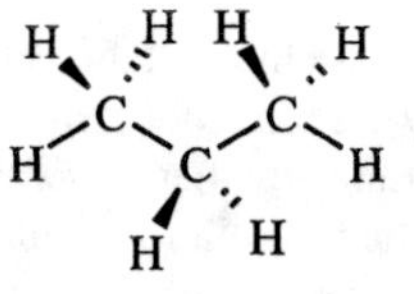

Perspective formula

Figure 1
The structural formula, condensed structural formulas, and perspective formula for propane (C_3H_8).

c. Branched alkanes. Branched alkanes are alkanes with a different connectivity of carbon atoms than straight-chain alkanes. The molecular formula for a branched chain alkane is the same as that of a straight-chain alkane (C_nH_{2n+2}), but the structure and properties are different. Branched-chain alkanes are isomers of straight-chain alkanes because they have the same molecular formula, but different structural formulas and different properties. Two of the different properties of branched-chain alkanes are lower boiling points and higher melting points as compared to straight-chain alkanes.

A branched-chain alkane has a substituent off of the longest carbon chain. If this substituent is a hydrocarbon, the name of the hydrocarbon will be the prefix to indicate the number of carbon atoms and the suffix –yl to indicate the alkyl group is an alkane minus one hydrogen atom.

The rules for naming branched-chain alkanes were developed by the International Union of Pure and Applied Chemists or IUPAC. These rules apply to the nomenclature of all organic molecules and should be committed to memory.

IUPAC Nomenclature.

1.) Identify the longest carbon chain and number each carbon atom of the longest chain starting at the end closest to a substituent. Apply the appropriate prefix to indicate the number of carbon atoms and use the suffix –ane.
2.) The alkyl group is be named next. Use a number to represent the carbon to which the substituent is bound. The number is separated from the name of the alkyl group by a hyphen.
3.) If more than one substituent is present, they are to be named in alphabetical order. If the substituents are the same, use a di- for two, tri- for three, and tetra- for four. Numbers should be separated from other numbers by commas, and numbers should be separated from text by hyphens.

Example: Give the name for the following branched-chain alkane.

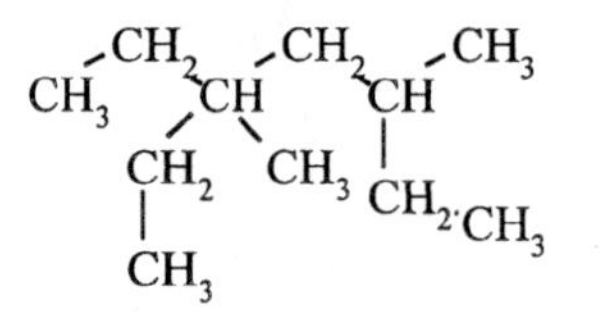

Solution:

1.) Determine the longest carbon chain. The longest carbon chain contains seven carbon atoms, so the parent name will be heptane.
2.) Number the carbon chain starting with the side that has the most substituents closest to the lowest number. The way this structure is drawn, number from left to right.
3.) There are methyl groups off of carbons numbered 3 and 5 and an ethyl group off of carbon number 3.
4.) Arrange the substituents in alphabetical order separating the number of the carbon they are attached to from the name of the substituent by a hyphen. For this molecule, the substituents would be listed as 3-ethyl-3,5-dimethyl.
5.) Put the name of the parent structure at the end of the substituent names and you're done. The IUPAC name for this compound is 3-ethyl-3,5-dimethylheptane.

d. Cycloalkanes. Cycloalkanes are hydrocarbons that have carbon atoms bonded together to form a ring. If you can imagine that every carbon atom forms four bonds, then it should make sense that if you bond two carbon atoms of an alkane together to make a ring, the formula for a cycloalkane will have two fewer hydrogen atoms then the formula for an alkane. The molecular formula for any cycloalkane is C_nH_{2n} where the n represents the number of carbon atoms. The nomenclature for cycloalkanes is the same as for alkanes except the prefix cyclo- is used. The additional facet to naming cycloalkanes is when there is more than one substituent present. The substituents may be geometric isomers of one another. If dashes or wedges are shown, you must specify whether the isomers are cis- or trans- to one another. The prefix cis- can be thought of as cis-same and trans-different. The cis-isomer has the substituents on the same side of the ring and the trans-isomer has the substituents on different sides of the ring.

e. Shapes of organic molecules. Chiral objects are nonsuperimposable mirror images of one another. If they were superimposable, they would be the same. In the same way that objects can be chiral, molecules can be chiral. For a molecule to be chiral, it must have an atom, which is usually carbon, with four different substituents off of it. Hydrogen can count as a substituent. Stereoisomers are molecules that differ in their three-dimensional arrangement of atoms in space. Stereoisomers are often chiral molecules. When two molecules are nonsuperimposable mirror images of one another, they are called enantiomers. Another important term to become familiar with is molecular conformation. Molecular conformations are the arrangement in space of atoms that results from rotation about single bonds. Many biological compounds are chiral and their activity or characteristics are a result of the chiral nature of the molecule.

f. Properties and reactions of alkanes. The physical properties of alkanes are easy to remember because you have experience with them every day of your life. The short chain alkanes exist as gases and as the chain length increases, they turn to liquids and eventually to solids. Some common examples are methane or natural gas, a gas, octane used to fuel your car, a liquid, and coal that you use to heat your grill, a solid or long chain hydrocarbon. How does this relate to physical properties? A gas exists as a gas because it lacks the intermolecular bonding necessary to gain a definite shape or occupy a definite volume. As the hydrocarbon chain grows, the London forces (Section 5.5), that are the only intermolecular forces holding hydrocarbons together, increase in influence. The result is that longer chain hydrocarbons have higher boiling points than shorter chain hydrocarbons. Branching of the hydrocarbon chain results in a decrease in the boiling point for alkanes. London forces work well for straight chain alkanes that can stack easily. The more branching in the hydrocarbon chain, the less packing and the lower the boiling point. It should be remembered that alkanes are nonpolar molecules and are thus not soluble in polar solvents like water. Another tidbit to take with you about alkanes is that they are the least reactive of the compounds you will encounter in the organic chemistry section. Alkanes will only undergo combustion or halogenation reactions. Both types of reaction have high activation energy barriers that must be overcome by heat or light. The combustion of an alkane is the reaction of the hydrocarbon to oxygen yielding carbon dioxide and water as products. The halogenation of an alkane will lead to the formation of an alkyl halide.

g. **Common functional groups of organic chemistry.** A functional group is an atom or group of atoms that gives a molecule its chemical and physical properties. Functional groups consist of unsaturated hydrocarbons and heteroatoms. Unsaturated hydrocarbons contain one or more double bonds and in some cases contain triple bonds. Oxygen, nitrogen, phosphorus, sulfur, and halogens are examples of heteroatoms. Heteroatoms are functional groups or parts of functional groups other than carbon or hydrogen.

Solutions to Odd Numbered Problems:

1. Organic chemists study the chemistry of carbon containing compounds.

3. a. Inorganic
   b. Organic
   c. Organic
   d. Inorganic

5. The compound is more likely to be inorganic. Inorganic compounds generally form ionic bonds that can be dissociated in an aqueous solution. Once the inorganic compounds dissociate, they form ions that are good electrolytes. Organic compounds are not readily soluble in water nor are they good electrolytes because they contain covalent bonds.

7. Organic compounds would be expected to have lower melting points. Think of a flower and a rock; which one is more rigid or solid in nature? The inorganic rock.

9. Organic

11. Carbon is in group 4A. It has four valence electrons and will need four more to satisfy the octet rule.

13. Hydrocarbons are organic compounds that contain only carbon and hydrogen atoms. Three common examples of hydrocarbons are methane, ethane, and propane.

15. Alkanes are saturated hydrocarbons containing carbon-carbon single bonds and carbon-hydrogen single bonds.

17. Methane has a tetrahedral shape.

19. Propane is the name of the alkane that contains three carbon atoms.

21.

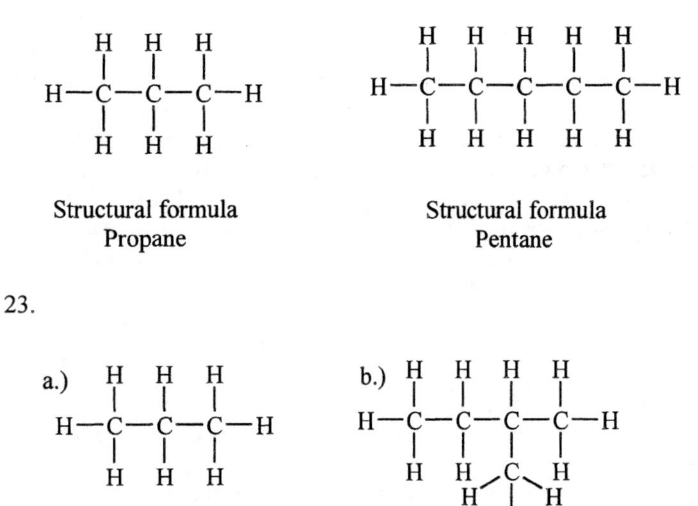

Structural formula
Propane

Structural formula
Pentane

23.

25. The most common condensed structural formula for ethane is CH_3CH_3. The most common condensed structural formula for nonane is $CH_3(CH_2)_7CH_3$ or $CH_3CH_2CH_2CH_2CH_2CH_2CH_2CH_2CH_3$.

27. The condensed structural formula for propane is $CH_3CH_2CH_3$.

29. Molecular models show organic molecules in three-dimensions where structural formulas only show organic molecules in two dimensions.

31. A wedge represents a bond to an atom that is pointing out of the two-dimensional plane of the drawing. A dashed line represents a bond to an atom that is pointing into the two-dimensional plane of the drawing.

33. a. Straight-chain alkane.
    b. Branched-chain alkane.
    c. Straight-chain alkane.

35. Isomers are molecules with the same molecular formulas but a different structural formulas and different properties.

37. a. Same compound.
    b. Isomer.
    c. Different compound.

39.

$CH_3CH_2CH_2CH_2CH_2CH_3$

CH_3
$CH_3CHCH_2CH_2CH_3$

CH_3
$CH_3CH_2CHCH_2CH_3$

CH_3
$CH_3CHCHCH_3$
CH_3

CH_3
$CH_3CCH_2CH_3$
CH_3

41. The name of an alkyl group is derived from the name of the corresponding alkane changing the suffix –ane to –yl. A molecule of propane can be converted to a propyl group by dropping the suffix –ane and replacing it with –yl. This change in name indicates that the propane molecule that has a formula C_3H_8 has lost a hydrogen atom to form a propyl group with a formula of C_3H_7.

43. a. The root name or parent name is the name given to the longest carbon chain.
    b. The prefix to the parent name consists of the substituents.
    c. The numbers prior to the substituents indicate the placement of the substituents on the longest carbon chain.

45. a. 4-ethyl-2-methylheptane
    b. 2,2,5-trimethylhexane
    c. 3,6-dimethylnonane
    d. 5-ethyl-2,2,5-trimethylheptane

47.

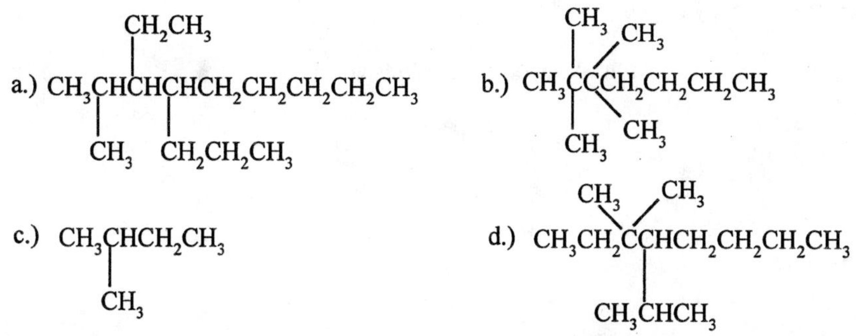

a.) CH_2CH_3 (above)
$CH_3CHCHCHCH_2CH_2CH_2CH_2CH_3$
CH_3 $CH_2CH_2CH_3$ (below)

b.) CH_3 CH_3 (above)
$CH_3CCCH_2CH_2CH_2CH_3$
CH_3 CH_3 (below)

c.) $CH_3CHCH_2CH_3$
CH_3 (below)

d.) CH_3 CH_3 (above)
$CH_3CH_2CCHCH_2CH_2CH_2CH_3$
CH_3CHCH_3 (below)

49. A cycloalkane is saturated hydrocarbon that has a carbon skeleton in the form of a ring.

51. a. methylcyclopentane
    b. ethylcyclobutane
    c. propylcyclohexane
    d. methylcyclooctane
    e. 1,1-diethylcyclopentane
    f. 1-ethyl-4-methylcyclohexane

53.

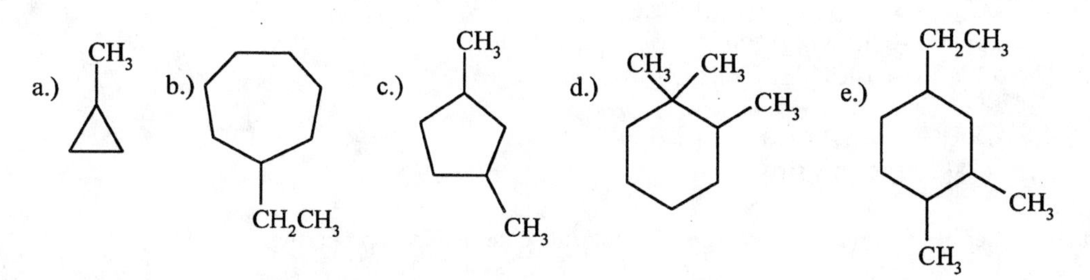

55. The substituent that is first alphabetically should occupy carbon number one. The name of this molecule should be 1-ethyl-2-methylcyclobutane.

57. The geometric isomer of trans-1,3-dimethylcyclopentane is cis-1,3-dimethylcyclopentane.

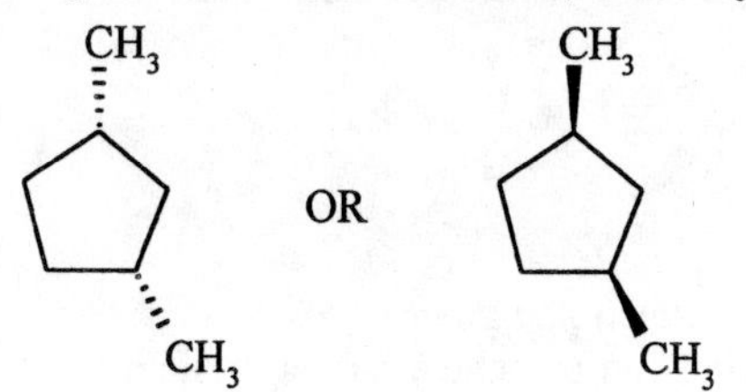

59. Cyclohexane is not planar because carbon atoms with four single bonds attached to them will have tetrahedral geometries. There is no way for an alkane to be planar because of the geometry about each carbon atom. The three-dimensional structure of cyclohexane is shown below in the chair form.

61. Chiral

63. b and c

65. The molecules are not hydrocarbons because they contain more than just carbon and hydrogen. The molecule in (b) has a chiral center because it has four different substituents attached. (a) is not a chiral carbon because two of the substituents are the same, hydrogen.

67. Carbon number three is chiral because it has four different substituents attached to it.

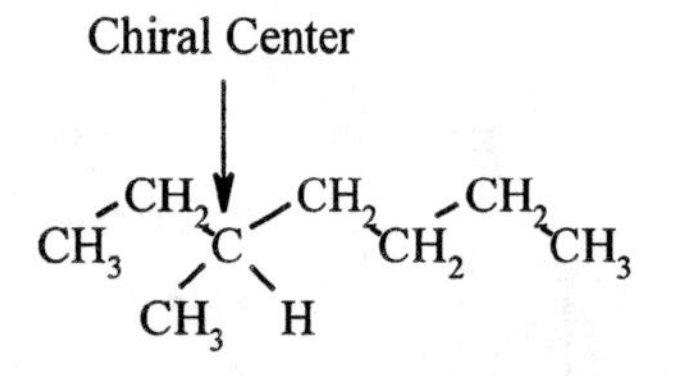

69. Free rotation is found around single covalent bonds.

71. Butane molecules will have more intermolecular interactions than methane molecules. The larger size of the hydrocarbon provides more surface area for interaction. This additional surface area leads to a greater influence by London dispersion forces. This trend continues with increasing chain length. As the hydrocarbon chain gets bigger, the interactions between nonpolar molecules increases.

73. Water is a polar solvent and will not dissolve alkanes or other nonpolar molecules.

75. The complete combustion of an alkane produces carbon dioxide and water.

77. a. $C_3H_8(g) + 5O_2(g) \rightarrow 3CO_2(g) + 4H_2O(g)$ + energy
    b. $2C_6H_{14}(l) + 19O_2(g) \rightarrow 12CO_2(g) + 14H_2O(g)$ + energy
    c. $2C_5H_{10}(l) + 15O_2(g) \rightarrow 10CO_2(g) + 10H_2O(g)$ + energy

79. a. $CH_3CH_3 + Br_2$ + light $\rightarrow CH_3CH_2Br + HBr$
    b. $CH_4 + Cl_2$ + light $\rightarrow CH_3Cl + HCl$

81. The structural theory states that the structure of a molecule determines its properties.

83. a. Alkynes = triple bond between two carbon atoms.
    b. Alkenes = double bond between two carbon atoms.
    c. Aromatic = alternating double and single bonds discussed in Section 9.5.

85. The structure of the functional group found in an amine is a nitrogen atom with three single bonds.

N

87. a. Aspirin contains an ester, carboxylic acid, and aromatic ring for functional groups.
    b. Tyrosine contains the functional groups phenol (aromatic alcohol), amine, and a carboxylic acid.

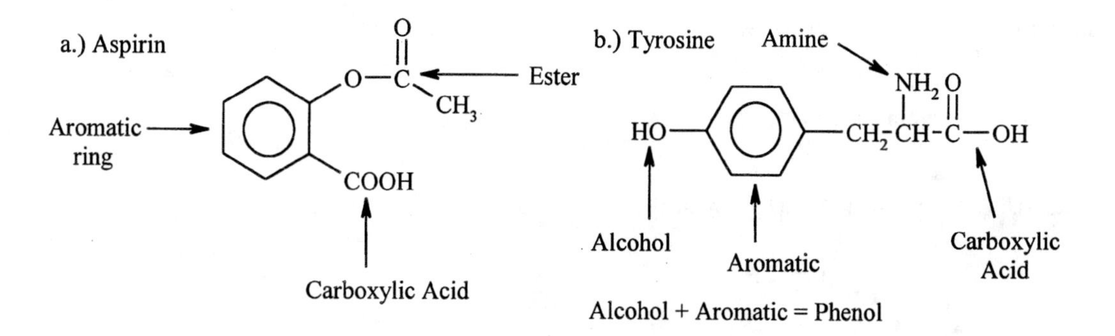

89. a. This problem assumes that you recall a couple of conversion factors. You must convert pounds of propane to moles of propane. The conversion factors for this conversion are 1 kg = 2.20 lb, and 1 kg = 1000 g. Start by determining the mass in grams of propane as follows.

20.0 lb propane $\times$ 1 kg/2.20 lb $\times$ 1000 g/1 kg = 9.10×10^3 g propane.

Use the molecular weight of propane (C_3H_8) to determine the number of moles of propane.

9.10×10^3 g C_3H_8 $\times$ 1 mol C_3H_8/44.11 g C_3H_8 = 2.06×10^2 mol C_3H_8

The problem tells you that each mole of propane releases 526.3 kcal when burned completely. Use this information to convert moles of propane to kcal heat generated by 2.06×10^2 mol C_3H_8.

2.06×10^2 mol C_3H_8 $\times$ 526.3 kcal/1 mol C_3H_8 = 1.09×10^5 kcal

To convert kcal to kJ, use the conversion factor 1 kcal = 4.184 kJ

1.09×10^5 kcal $\times$ 4.184 kJ/1 kcal = 4.56×10^5 kJ

b. $C_3H_8(g) + 5O_2(g) \rightarrow 3CO_2(g) + 4H_2O(g)$ + energy

c. You will need to use the information obtained in parts a and b to answer this question. In part a of the problem, you learned that 20.0 lb of propane = 2.06×10^2 mol C_3H_8, and in part b you determined that for every 1 mole of propane that reacts, 3 moles of carbon dioxide are produced. Use this information to determine the moles of carbon dioxide produced by 2.06×10^2 mol C_3H_8.

2.06×10^2 mol C_3H_8 $\times$ 3 mol CO_2/1 mol C_3H_8 = 6.18×10^2 mol CO_2

Use the molecular weight of carbon dioxide to convert from moles to grams.

6.18×10^2 mol CO_2 $\times$ 44.01 g CO_2/1 mol CO_2 = 2.72×10^4 g CO_2

Use the conversion factors from part a of the problem to convert from grams to pounds.

2.72×10^4 g CO_2 $\times$ 1 kg/1000 g $\times$ 2.20 lb/1 kg = 5.98×10^1 lb CO_2

The answer is 20.0 lb of propane will produce 59.8 lbs of carbon dioxide upon complete combustion.

Sample Test:

1. Which of the following compounds is organic?
   a. $HCHO_2$ b. H_2O c. H_2SO_4 d. NaOH

2. How many hydrogen atoms will a molecule of octane contain?
   a. 8 b. 16 c. 18 d. 24

3. What do all hydrocarbons have in common?
   a. They combust.
   b. They are often found in nonliving matter.
   c. They contain hydrogen and carbon.
   d. More than one correct response.

4. Which is the correct perspective formula for ethane?

a.)
```
    H  H
    |  |
 H—C—C—H
    |  |
    H  H
```
b.)
```
 H\      /H
   C—C
H'' |  | ''H
    H  H
```
c.) CH_3-CH_3

d.) C_2H_6

5. How many isomers are there for pentane?
   a. 1 b. 2 c. 3 d. 5

6. How many hydrogen atoms will a butyl group have?
   a. 4 b. 8 c. 9 d. 10

7. What is the IUPAC name of the branched-chain alkane drawn below?

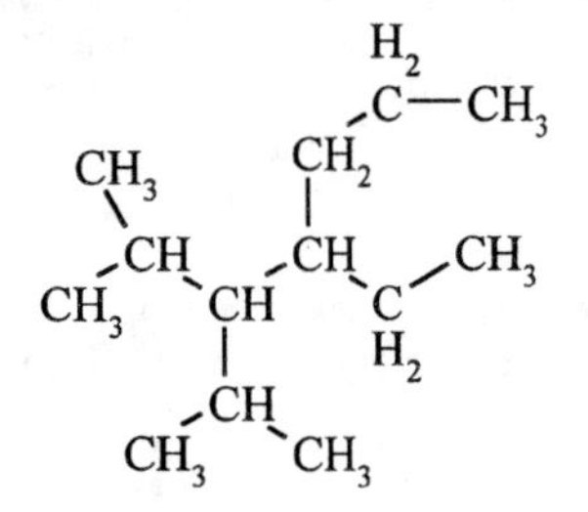

   a. 4-ethyl-3-isopropyl-2-methylheptane
   b. 5-methyl-3,4-dipropylhexane
   c. 3,4-dipropyl-5-methylheptane
   d. 2-methyl-3,4-dipropylhexane

8. What is the correct structure for 2,2-dimethylbutane?

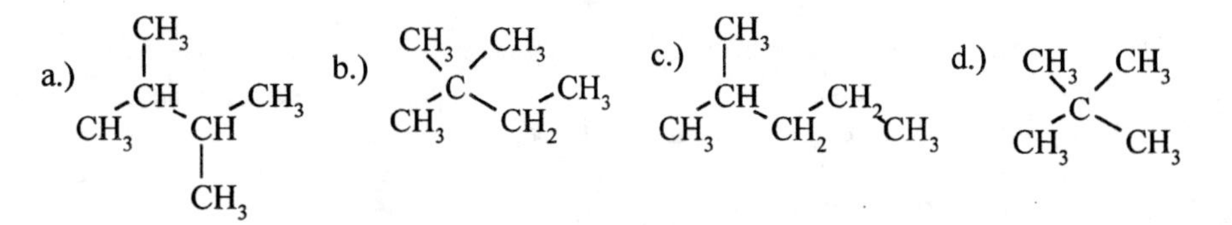

9. What is the correct name for 1-ethyl-2-methyl-3-ethylcyclohexane?
   a. 1-methyl-2,6-diethylcyclohexane
   b. 2-methyl-1,3-diethylcyclohexane
   c. 1,3-diethyl-2-methylcyclopentane
   d. 1,3-diethyl-2-methylcyclohexane

10. What is the name of the following molecule?

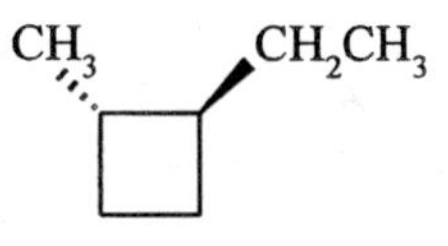

   a. ethylmethylcyclobutane
   b. 1-ethyl-2-methylcyclobutane
   c. cis-1-ethyl-2-methylcyclobutane
   d. trans-1-ethyl-2-methylcyclobutane

11. In the compound 2,2-diethyl-3-methylheptane, what number carbon(s) is/are the chiral center(s)?
   a. 2 b. 2 and 3 c. 3 d. There are no chiral centers in the molecule.

12. What relationship exists between cis-1,2-dimethylcyclobutane and trans-1,2-dimethylcyclobutane?
   a. They are enantiomers.
   b. They have the same molecular conformation.
   c. They are stereoisomers.
   d. They are superimposable mirror images of one another.

13. How many molecules of oxygen will react in the complete combustion of butane?
    a. 1 b. 7 c. 10 d. 13

14. Based on your newly acquired knowledge of alkanes, can you makes sense of why many large cities are switching to natural gas (methane) powered buses rather than petroleum gas (mixture of longer chain hydrocarbons) powered buses?
    a. Methane has a higher boiling point so it will undergo more complete combustion.
    b. Methane has a lower boiling point so it will undergo more complete combustion.
    c. Longer chain hydrocarbons will have lower boiling points so they will not undergo complete combustion.
    d. More than one correct answer.

15. The complete halogenation of methane in the presence of bromine will give which of the following end products?
    a. CH_3Br b. CH_2Br_2 c. $CHBr_3$ d. CBr_4

16. Which of the following solvents would hexane dissolve in?
    a. Decane b. Water c. Sulfuric acid d. Sodium Hydroxide

17. The addition of chlorine to an alkane will give what class of compound as a product?
    a. Alcohol b. Alkyl halide c. Amine d. Ester

18. What element must be present in a thiol functional group?
    a. Sulfur b. Nitrogen c. Oxygen d. Carbon

19. How many oxygen atoms will be involved in the complete combustion of 25.0 g of methane (CH_4)?
    a. 5.18×10^{-24} oxygen atoms
    b. 9.39×10^{23} oxygen atoms
    c. 3.76×10^{24} oxygen atoms
    d. 1.89×10^{24} oxygen atoms

20. How many functional groups are there in the molecular structure of cocaine?

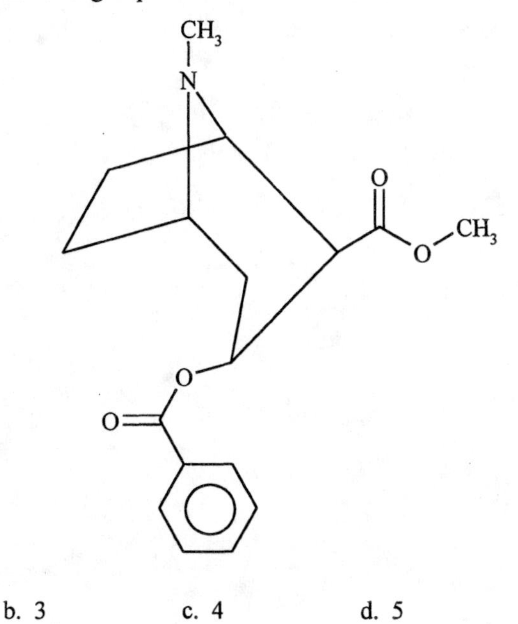

    a. 2 b. 3 c. 4 d. 5

Answers to Sample Test:

1. a
2. c
3. d
4. b
5. c
6. c
7. a
8. b
9. d
10. d
11. c
12. c
13. d
14. b
15. d
16. a
17. b
18. a
19. c
20. c

Chapter 9: Unsaturated Hydrocarbons

Chapter Objectives:

1. To recognize that alkenes are unsaturated hydrocarbons.
2. To learn the nomenclature and structural formulas for alkenes.
3. To learn about compounds containing carbon-carbon triple bonds.
4. To understand the physical and chemical interaction of alkenes and alkynes with other molecules.
5. To learn about aromatic hydrocarbons.
6. To understand the physical and chemical properties of aromatic hydrocarbons.

Key Terms: The key terms are listed in the order in which they are encountered in the chapter.

Unsaturated hydrocarbon. A compound that contains carbon and hydrogen with fewer than the maximum number of hydrogen atoms.

Alkenes. Unsaturated hydrocarbons that contain one or more carbon-carbon double bonds that are the functional groups of the molecule.

Terminal double bond. A double bond at either end of the longest carbon chain in an alkene.

Alkynes. Unsaturated hydrocarbons containing the carbon-carbon triple bond functional group that follow the molecular formula C_nH_{2n-2}.

Addition reaction. The addition of a reactant to each of the carbon atoms of a double bond that converts the double bond to a single bond.

Hydrogenation. The addition of a molecule of hydrogen across a double bond to make a single bond in the presence of a metal catalyst like platinum or nickel.

Dehydrogenation. The removal of a molecule of hydrogen from two adjacent carbon atoms in a molecule to make a double bond.

Halogenation. The addition of two halogen atoms of a halogen molecule to the two carbon atoms of a double bond to make an alkyl halide.

Hydration. The addition of a molecule of water to a double bond to make an alcohol under acidic conditions.

Markovnikov's rule. When an asymmetrical, hydrogen-containing reagent adds to a carbon-carbon double bond, the carbon atom of the double bond that has more hydrogen atoms gets the hydrogen atom of the reagent.

Hydrohalogenation. The addition of a hydrohalogen or hydrogen halide to the carbon atoms of a double bond to produce an alkyl halide according to Markovnikov's rule.

Hydroxylation. An oxidation reaction that produces a molecule containing a hydroxyl group on both carbon atoms of a double bond.

Oxidative cleavage. Under strong oxidative conditions, a double bond may be cleaved and a carbon-oxygen double bond formed in its place.

Polymerization. The combination of monomers to make polymers.

Monomer. A small molecular building-block.

Polymer. A bunch of monomers joined together.

Aromatic hydrocarbons. A compound that contains one or more benzene rings as its functional group.

Aromatic. The presence of an aromatic ring (benzene molecule) in a compound.

Delocalized electrons. Electrons that are shared equally throughout the atoms of a molecule that has alternating single and double bonds giving the semblance of a bond-and-a-half.

Benzene. The simplest of aromatic hydrocarbons with the formula C_6H_6.

Toluene. The accepted name for a common aromatic compound that consists of a methyl group attached to benzene.

Aniline. The accepted name for an aromatic compound that consists of an amino ($-NH_2$) group attached to an aromatic ring.

Ortho-. The term used to describe two substituents located in the 1 and 2 position of an aromatic ring.

Meta-. The term used to describe two substituents located in the 1 and 3 position of an aromatic ring.

Para-. The term used to describe two substituents located in the 1 and 4 position of an aromatic ring.

Polycyclic. Two or more rings fused together.

Phenyl group. An aromatic ring minus a hydrogen atom, so the formula is C_6H_5.

Naphthalene. A polycyclic aromatic compound containing two fused aromatic rings.

Anthracene. A polycyclic aromatic compound containing three fused aromatic rings.

Substitution reaction. A chemical reaction involving an aromatic ring where a hydrogen atom on the ring will be replaced by part of a reactant molecule.

Alkylation. The reaction of a hydrocarbon where a alkyl group (hydrocarbon minus a hydrogen atom) is added or substituted to another reactant.

Nitration. A reaction involving nitric acid where a nitro (NO_2) group is transferred to a reactant.

Sulfonation. A reaction involving the addition or substitution of a sulfonic acid group (SO_3H) to another reactant.

Chapter 9 Objective Details:

a. Alkenes. Alkenes are unsaturated hydrocarbons where the double bonds are the functional groups. The formula for an alkene is C_nH_{2n}, where n is the number of carbon atoms. The formula decreases by 2 hydrogen atoms for each additional double bond. If you remember, the formula for alkanes was C_nH_{2n+2}. When a double bond is added, the formula changes to C_nH_{2n}. A second double bond in a molecule will have the formula C_nH_{2n-2}. Naming alkenes is very much the same as naming alkanes. The longest carbon chain containing the double bond is the parent chain. The number of carbon atoms in the parent chain establishes the prefix eth-, prop-, but-, etc. The suffix changes from –ane in alkanes to –ene for alkenes. The double bond is the functional group of the molecule, so you must specify its position with a number. Number the parent chain such that the lowest number is used for one of the carbons of the double bond. This number precedes the name of the molecule. If other substituents are present, they are listed in alphabetical order and numbered by their position on the parent chain. Remember to include the prefixes di-, tri-, and tetra- for more than one functional group and separate numbers from number by a comma and numbers from words by a hyphen.

b. Alkene structure. Alkenes have double bonds that do not have freedom of rotation as was observed with alkanes. This rigid structure about a double bond means that cis- and trans- isomers exist for many alkenes. For an alkene to have a cis- or trans- orientation, there cannot be two of the same atom or group attached to a carbon of the double bond. This is the reason that molecules with terminal double bonds do not have different isomers. An alkene that is cis- will have the two smallest groups in one plane of the double bond and the two largest groups in the other plane of the double bond. Plane means that the double bond makes the molecule planar or flat. The trans- orientation has the large groups in opposite planes across the double bond and the smaller groups on opposite sides of the double bond. Perhaps an example is in order.

Example: Compare the structures of 2-methyl-2-butene, cis-2-butene, and trans-2-butene.

Solution:

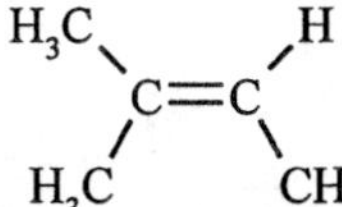

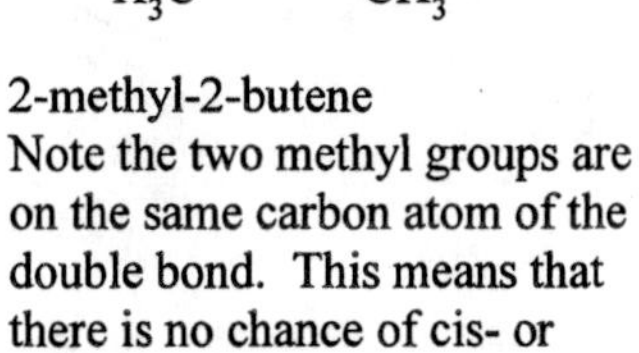
2-methyl-2-butene
Note the two methyl groups are on the same carbon atom of the double bond. This means that there is no chance of cis- or trans- isomers.

H H
C=C
H3C CH3

Cis-2-butene
The two large groups (methyls) are on the same side of the double bond and the two small groups (H atoms) are on the same side of the double bond, therefore the molecule is the cis-isomer.

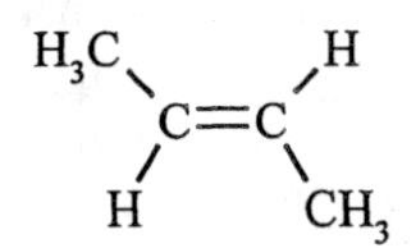

Trans-2-butene
The two large groups are opposite one another across the double bond and the two small groups are also opposite one another making this trans.

c. Alkynes. Alkynes are unsaturated hydrocarbons that contain a carbon-carbon triple bond. The two degrees of unsaturation in an alkyne lead to the general formula for alkynes as C_nH_{2n-2}. A triple bond between two carbon atoms gives alkynes a linear geometry ruling out the possibility of cis- and trans-isomers provided the triple bond is the only functional group. Alkanes end in the suffix –ane, alkenes end in –ene, and alkynes end in –yne. The rules that apply to the nomenclature of alkenes also apply to alkynes.

d. Properties and reactions of alkenes and alkynes. Alkenes and alkynes are similar to alkanes in that they are nonpolar hydrocarbons held together by nothing more than London dispersion forces. The same trend of increasing boiling point with increasing carbon chain length still holds true. It is also true that neither alkenes nor alkynes are soluble in polar solvents.

For the context of this book, alkenes can undergo four types of addition reactions, two types of oxidation reactions, and polymerization reactions. An addition reaction is just that; a reactant adds across the double bond to make a single bond. Types of reactants that can add across the double bond include a hydrogen molecule (hydrogenation), a halogen molecule (halogenation), a water molecule (hydration), and a hydrogen halide (hydrohalogenation). There are two types of reactants listed here. Molecular hydrogen and halogens are symmetrical, while water and hydrogen halides are asymmetrical. The symmetrical molecules (H_2, Cl_2, Br_2, I_2), will add one atom to each carbon of a double bond to create a carbon-carbon single bond. The asymmetrical reactants (H-OH, H-Cl, H-Br, H-I) will follow Markovnikov's rule. The rule states that the hydrogen atom of the reactant will add to the side of the double bond that contains the most hydrogen atoms. The oxidation reactions of alkenes are termed hydroxylation (for the addition of hydroxyl groups), and oxidative cleavage. The hydroxylation reactions are in the presence of a mild oxidizing agent and result in the formation of a diol. A diol is a hydrocarbon containing two alcohol functional groups. The hydroxyl groups add to each carbon of a carbon-carbon double bond during the course of the reaction. Oxidative cleavage occurs under severe oxidizing conditions and results in the cleavage of the carbon-carbon double bond. Where the carbon-carbon double bond used to be a carbon-oxygen double bond appears. Polymerization is the last class of reaction that deals with the combining of alkene monomer units to make much larger molecules called polymers. Before moving to reactions of alkynes, it is worth noting that hydrogenation reactions require a metal catalyst like platinum or nickel, and hydration reactions require acidic conditions.

Alkynes react similar to alkenes with the exception that alkynes have a triple bond that is equivalent to two double bonds. The reaction of an alkyne with hydrogen will require two moles of molecular hydrogen to reduce the triple bond to a single bond. For addition of asymmetrical reactants across the double bond, Markovnikov's rule still applies.

e. Aromatic compounds. Aromatic compounds are hydrocarbons that contain delocalized electrons. The electrons of alternating double and single bonds are shared equally between all of the carbon atoms of a six-member ring. The simplest of all aromatic compounds and the basic building block for them is benzene (C_6H_6). Benzene is the functional group of all aromatic compounds. Aromatics as they are referred to have many common names that are accepted. Toluene, aniline, and the prefixes ortho-, meta-, and para- are considered accepted nomenclature for aromatics. An aromatic ring that is a substituent or functional group for another molecule is referred to as a phenyl group. There are a few compounds that you should be familiar with and those are summarized below.

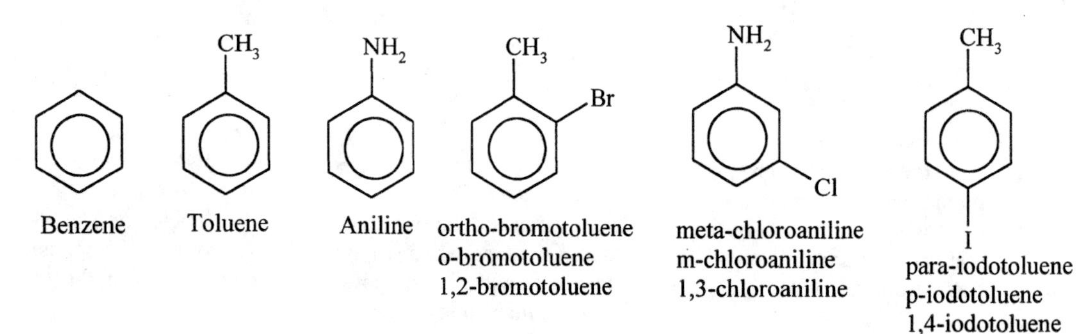

f. Properties and reactions of aromatic hydrocarbons. The key term here is again hydrocarbon. Hydrocarbons are nonpolar by definition and are thus not soluble in organic solvents. The forces that hold aromatic hydrocarbons together are London dispersive forces. Aromatic compounds contain delocalized electrons that are far less reactive and more stable than normal double bonds. The delocalized nature of the bonds does not get interrupted through the course of a chemical reaction. Alkenes were prone to addition reaction and aromatic hydrocarbons are prone to substitution reactions. You will be asked to remember four substitution reactions including alkylation, halogenation, nitration, and sulfonation. The alkylation reaction involves the reaction of an alkyl halide with an aromatic hydrocarbon in the presence of the catalyst aluminum chloride to produce the alkylated aromatic compound. The halogenation reaction involves the reaction of a halogen molecule with an aromatic ring in the presence of a metal halide catalyst to produce the halogenated aromatic ring. The nitration reaction involves the reaction of an aromatic hydrocarbon with nitric acid in the presence of a sulfuric acid as the catalyst to produce the nitrated aromatic ring. The sulfonation reaction involves the reaction of an aromatic hydrocarbon in the presence of sulfuric acid to produce the sulfonic acid aromatic ring. These reactions are summarized for your convenience.

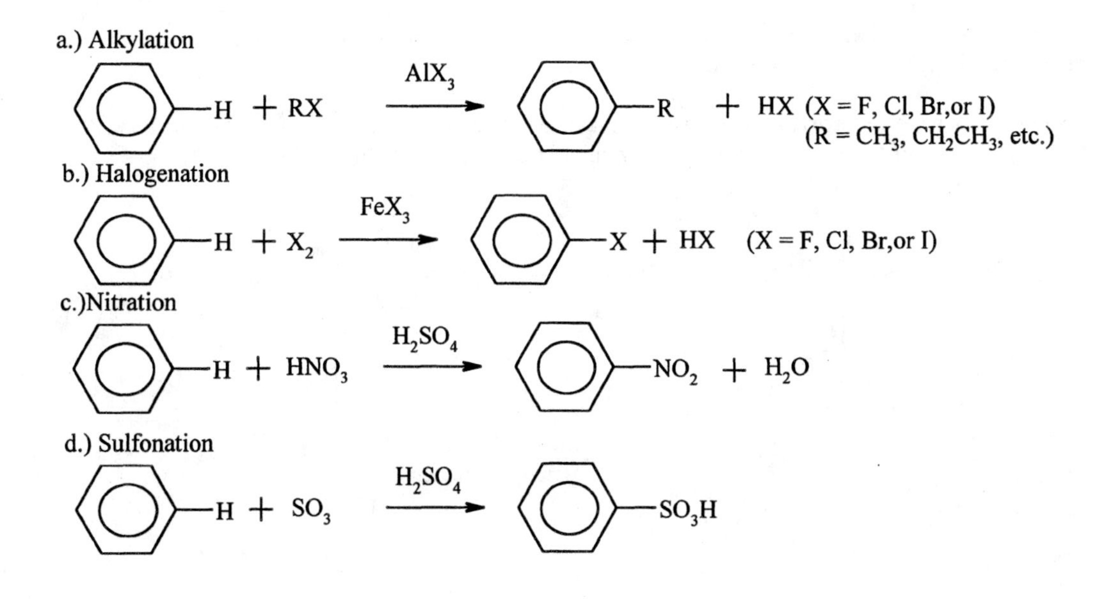

Solutions to Odd Numbered Problems:

1. An unsaturated hydrocarbon is a compound that contains carbon and hydrogen with fewer than the maximum number of hydrogen atoms.

3. The three classes of hydrocarbons that are unsaturated are the alkenes, alkynes, and aromatics.

5. a. The principle source of alkenes is petroleum refining.
   b. Alkenes are used to make plastics, fibers, alcohol, antifreeze, and vinegar.

7. Ethene

9.

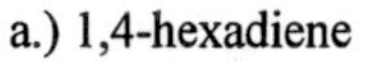

H H
H_2C=C C C=C CH_3
H_2 H

b.) Cycloheptene

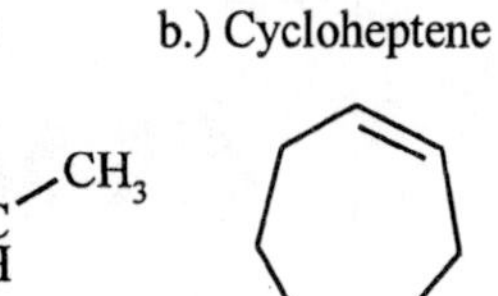

c.) Propylene

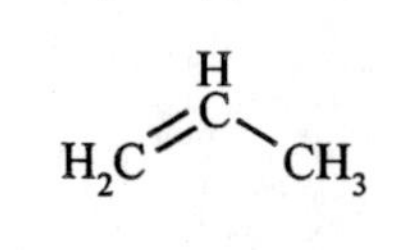

d.) 3-heptene

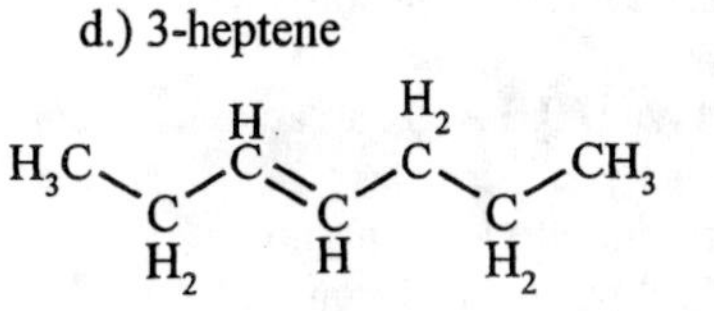

11. a. 3-heptene
    b. 4,5-dimethyl-1-hexene
    c. 2-methyl-3-propylcyclohexene

13.

a.) Cyclohexene b.) 3-methylcyclopentene

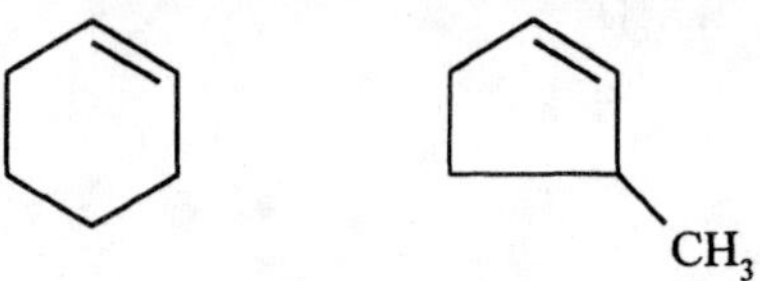

15. Free rotation can occur around single covalent bonds.

17. Cis- and trans-isomers can exist across carbon-carbon double bonds. An alkene that is cis- will have the two smallest groups in one plane of the double bond and the two largest groups in the other plane of the double

bond. The trans- orientation has the large groups in opposite planes across the double bond and the smaller groups on opposite sides of the double bond.

19. 1-pentene will not have geometric isomers because the double bond is terminal. Terminal double bonds contain two hydrogen atoms off of the same carbon making cis- or trans-isomers not appropriate.

21.

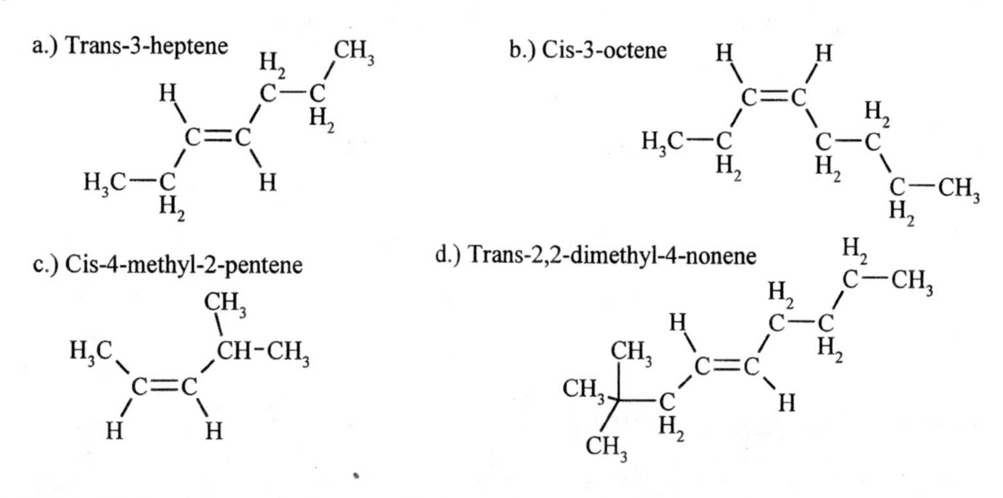

23. The double bond starts at carbon number 2, so it should be called 2-butene.

25. a. Trans-9-methyl-4-decene
    b. Cis-4-methyl-2-pentene
    c. Trans-3,4-dimethyl-3-heptene

27. The most common alkyne is ethyne also known as acetylene. The structural formula for ethyne is: H-C≡C-H. The principle use of ethyne is in oxyacetylene torches.

29. The compound cis-2-butyne does not exist because cis- and trans-isomers of alkynes are not possible. The carbon-carbon triple bond is linear and does not allow for geometric isomers.

31. a. Ethyne (IUPAC) and acetylene (common)
    b. 2-hexyne
    c. 4,4-dimethyl-2-pentyne
    d. 2,6,8-trimethyl-4-nonyne

33.

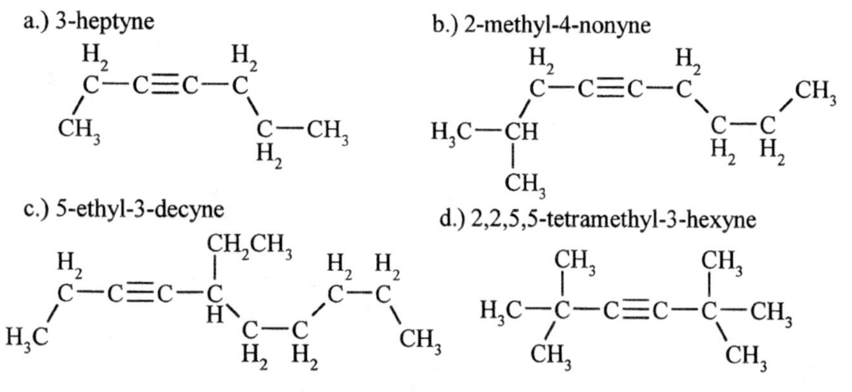

35. Alkenes and alkynes are held together by London dispersion forces. As the size of the hydrocarbon chain increases, the strength of the London forces increases.

37. Alkenes can react by addition, oxidation, or polymerization.

39. a. $CH_3CH_2CH_2CH_3$
    b. $CH_3CH(CH_3)CH_2CH_3$
    c. $CH_3CH_2CH_2CH_3$

41. Halogen atoms become bonded to the carbon atoms of the carbon-carbon double bond during a halogenation reaction.

43.

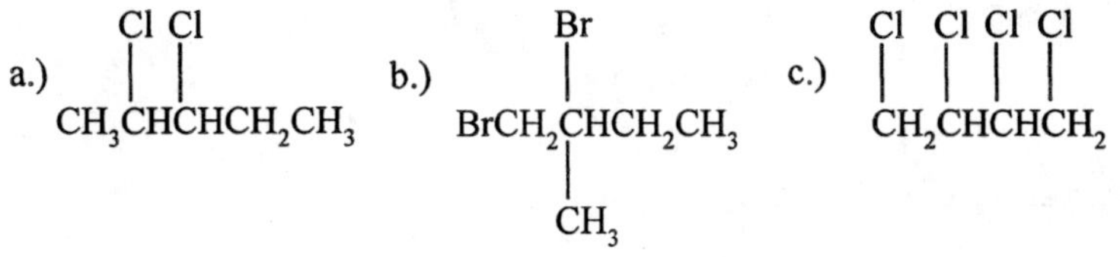

a.) $CH_3CH(Cl)CH(Cl)CH_2CH_3$

b.) $BrCH_2C(Br)(CH_3)CH_2CH_3$

c.) $CH_2(Cl)CH(Cl)CH(Cl)CH_2(Cl)$

45. Markovnikov's rule states that when an asymmetrical, hydrogen-containing reagent adds to a carbon-carbon double bond, the carbon atom of the double bond that has more hydrogen atoms gets the hydrogen atom of the reagent. Markovnikov's rule would not apply to the hydration of ethene because both carbon atoms of the double bond have the same number of hydrogen atoms.

47.

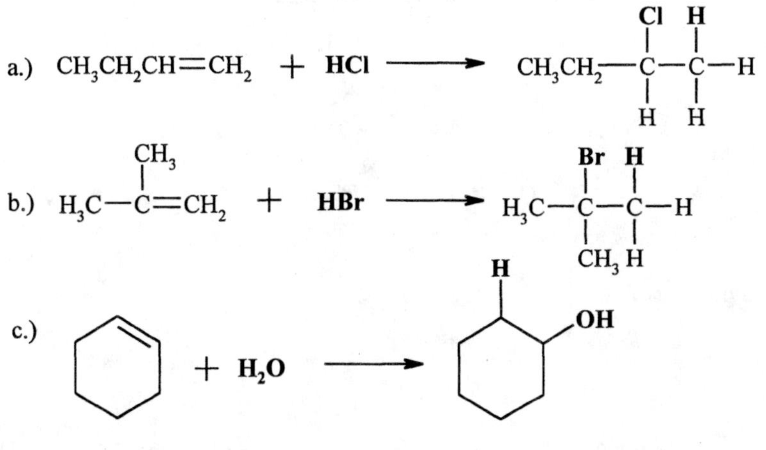

a.) $CH_3CH_2CH{=}CH_2$ + **HCl** ⟶ $CH_3CH_2{-}C(Cl)(H){-}C(H)(H){-}H$

b.) $H_3C{-}C(CH_3){=}CH_2$ + **HBr** ⟶ $H_3C{-}C(Br)(CH_3){-}C(H)(H){-}H$

c.) cyclohexene + **H_2O** ⟶ cyclohexanol (**H** and **OH** on adjacent ring carbons)

49.

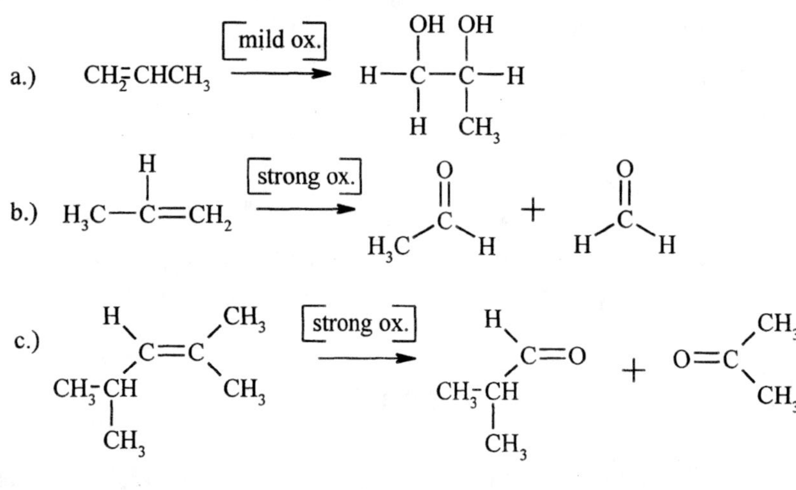

51. $CH_2{=}CH_2 + CH_2{=}CH_2 + CH_2{=}CH_2 + \ldots \rightarrow -CH_2\text{-}CH_2\text{-}CH_2\text{-}CH_2\text{-}CH_2\text{-}CH_2\text{-}$

53. A compound is considered to be aromatic if an aromatic ring is present.

55. Delocalized electrons are electrons that are shared equally among the carbon atoms of an aromatic compound. Aromatic hydrocarbons always contain delocalized electrons.

57.

a.) o-ethyltoluene b.) m-propyltoluene c.) 1,3-diethylbenzene

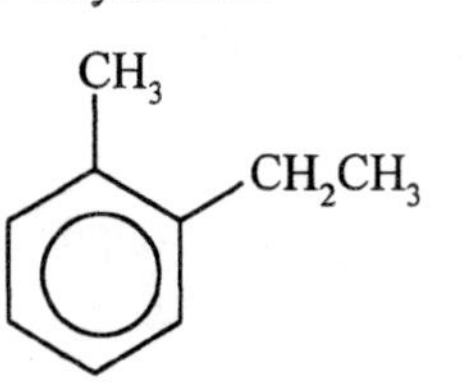

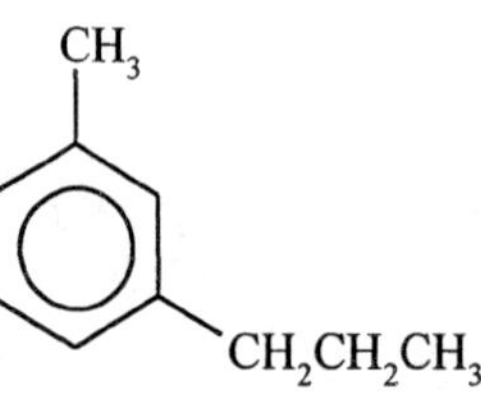

59. a. Propylbenzene
    b. 1,2-diethylbenzene or o-diethylbenzene
    c. 2,4-diethyltoluene

61.

a.) Chlorobenzene b.) Bromobenzene c.) p-bromotoluene d.) 1,2-difluorobenzene e.) 1-bromo-2,4-dichlorobenzene

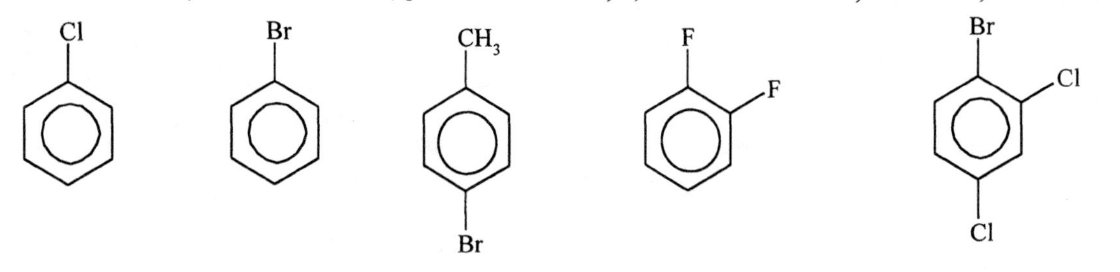

63. a. 1-bromo-3-fluorobenzene or m-bromofluorobenzene
    b. 2-nitrotoluene or o-nitrotoluene
    c. 1,2-dibromo-3-nitrobenzene

65. Polycyclic compounds have two or more rings fused together. Two common polycyclic aromatic compounds are naphthalene and anthracene.

67. Aromatic hydrocarbons behave as all other hydrocarbons in that they are soluble in nonpolar solvents and insoluble in polar solvents.

69.

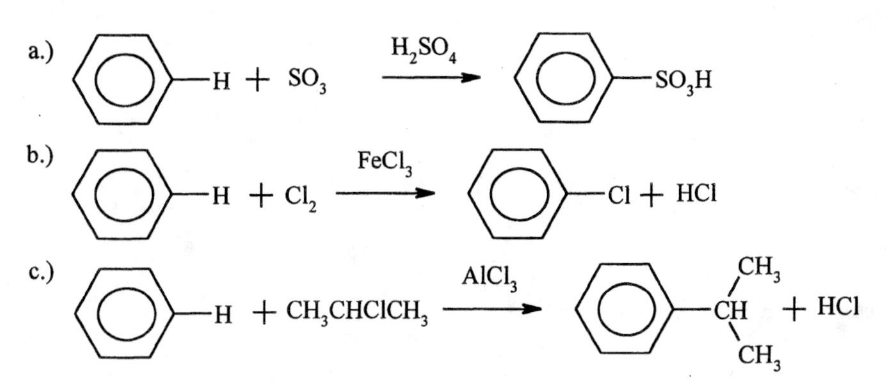

71. For a step-by-step solution to a problem of this nature, refer to Chapter 8, problem number 89, on page 83 of this manual.
a. $2\ C_2H_2 + 5\ O_2 \rightarrow 4\ CO_2 + 2\ H_2O$ + energy
b. 100.0 g of C_2H_2 is the equivalent of 3.84 mol of C_2H_2. Since each mol of C_2H_2 releases 312 kcal of energy, the combustion of 3.84 mol of C_2H_2 will liberate 1.20×10^3 kcal of energy. To convert to kJ, use the conversion factor 4.184 kJ = 1 kcal. 100.0 g of C_2H_4 will liberate 5.02×10^3 kJ of energy. For a step-by-step solution to a problem of this nature, refer to Chapter 8 problem number 89 on page 83 of this manual.
c. From the balanced chemical equation in step a, you know that for every two moles of acetylene that react, five moles of oxygen are required. In Chapter 5, Section 5.4 you learned that one mole of a gas will occupy a volume of 22.4 L. A combination of this information is needed to solve this problem.

$3.84 \text{ mol } C_2H_2 \times (5 \text{ mol } O_2/2 \text{ mol } C_2H_4) \times (22.4 \text{ L } O_2/1 \text{ mol } O_2) = 215 \text{ L } O_2$

Sample Test:

1. Which of the following molecules is a alkene with one double bond?
   a. C_3H_8 b. C_6H_{10} c. $C_{14}H_{28}$ d. $C_{15}H_{26}$

2. Name the following alkene.

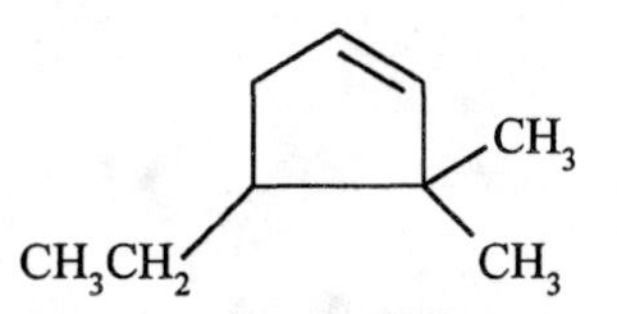

   a. 3,3-dimethyl-4-ethyl-1-pentene
   b. 3,3-dimethyl-4-ethyl-1-cyclopentene
   c. 4-ethyl-3,3-dimethyl-1-cyclopentene
   d. 4-ethyl-3,3-dimethylcyclopentene

3. What is the name for the following alkene?

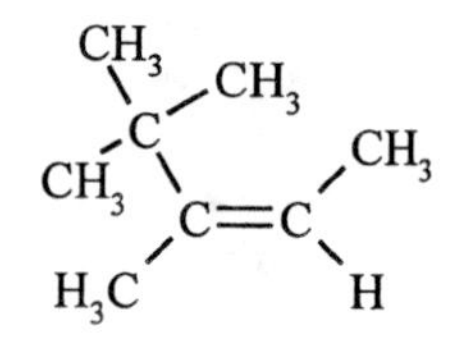

a. 2,2,2-trimethyl-2-butene
b. 3,4,4-trimethyl-2-pentene
c. Cis-3,4,4-trimethyl-2-pentene
d. 2,2,3-trimethyl-3-pentene

4. Which is the correct IUPAC name for Trans-2-ethyl-2-pentene?
   a. Cis-2-ethyl-2-pentene
   b. Cis-3-methyl-3-hexene
   c. Cis-3-methyl-3-pentene
   d. Trans-3-methyl-3-hexene

5. What is the correct structure of 1,3-dimethylcyclopentene?

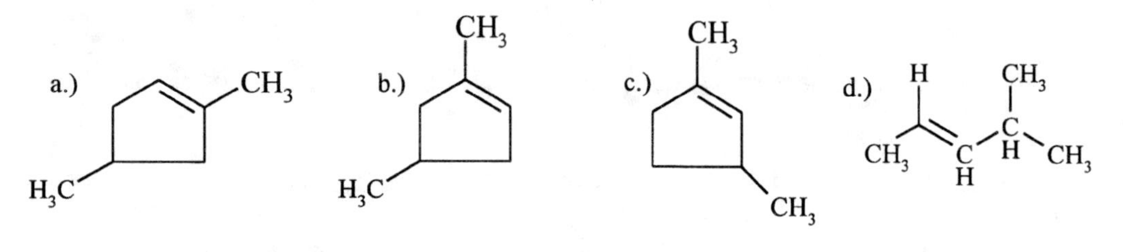

6. What's the formula for 2-pentyne?
   a. C_5H_{12} b. C_5H_{10} c. C_5H_{14} d. C_5H_8

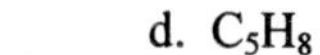

7. What is the IUPAC name of the alkyne drawn below?

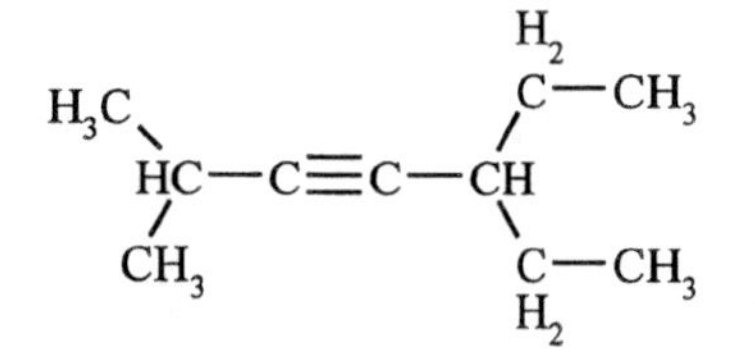

a. 5-ethyl-2-methyl-3-heptyne
b. 3-ethyl-6-methyl-4-heptyne
c. 5,5-diethyl-2-methyl-3-pentyne
d. Cis-3-ethyl-6-methyl-4-heptyne

8. What is the correct structure 5-methyl-2-hexyne?

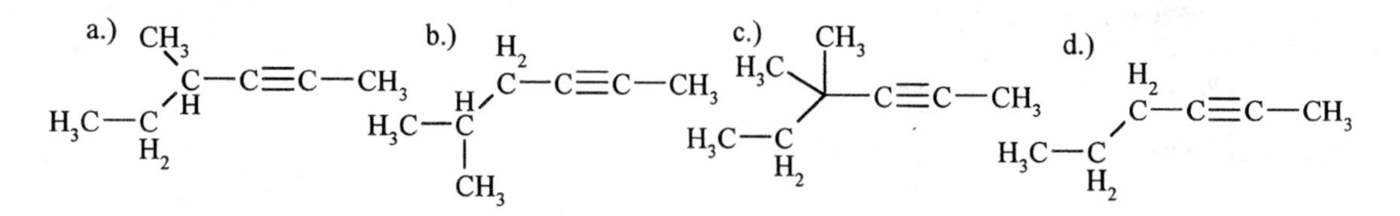

9. What will be the major product in the following reaction?

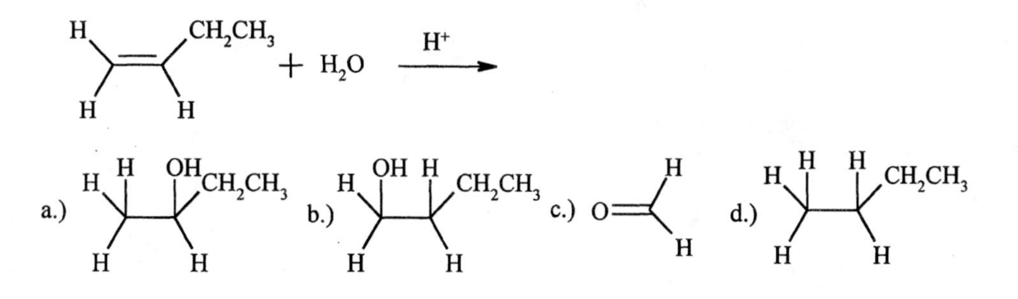

10. What is the name of the molecule produced by the following reaction?

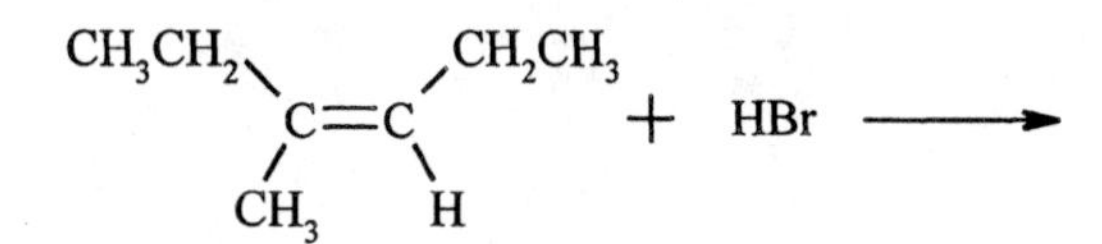

a. Cis-3-methyl-3-hexene
b. Cis-3-methyl-3-hexane
c. 3-methyl-3-hexane
d. 3-methylhexane

11. What will be the major product in the following reaction?

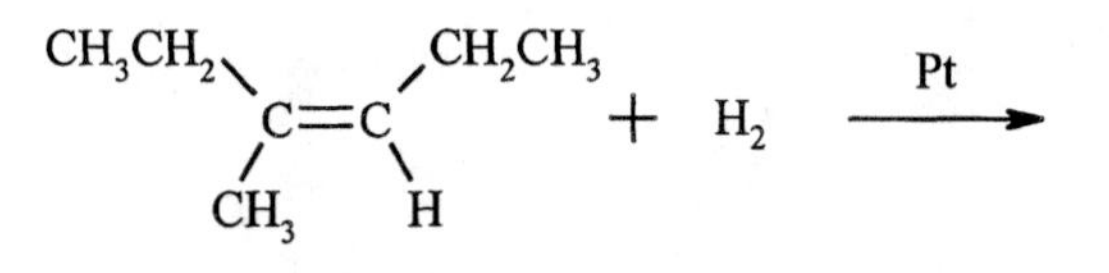

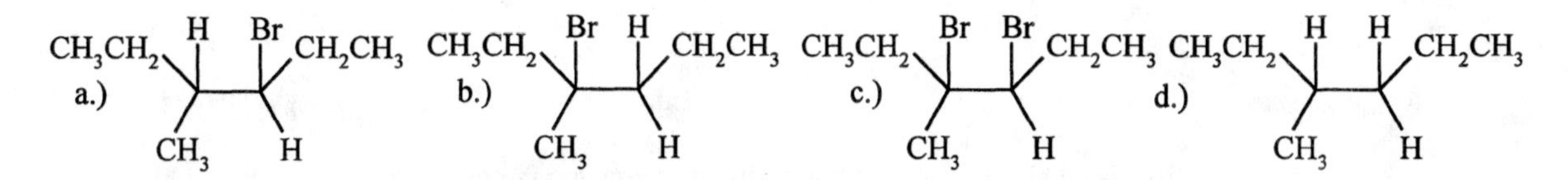

12. Under what conditions will an alkene be broken down into an aldehyde or a ketone?
a. Polymerization b. Mild Oxidation c. Strong oxidation d. Addition

13. Oxidation of what alkene would give you the following product?

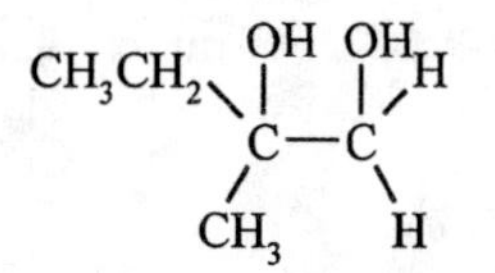

a. 2-methyl-1-butene
b. 2-methyl-2-butene
c. Cis-2-methyl-1-butene
d. Trans-2-methyl-2-butene

14. How many moles of hydrogen molecules would be necessary to completely reduce 1.5 moles of acetylene?
a. 1.5 b. 3.0 c. 4.5 d. 6.0

15. What is the difference between benzene and a phenyl group?
a. Benzene is an aromatic ring and a phenyl group is not.

b. Benzene is the parent structure and a phenyl group is benzene minus a hydrogen atom.
c. A phenyl group is any aromatic hydrocarbon and benzene is a specific aromatic ring.
d. There is no difference between the two.

16. What is the name of the following aromatic hydrocarbon?

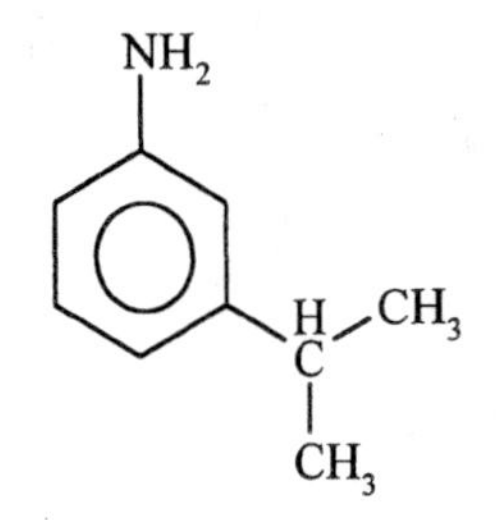

a. 1,3-propylaniline
b. p-isopropylaniline
c. o-isoproplyaniline
d. m-isopropylaniline

17. What is the name used for the compound p-ethylmethylbenzene?
a. 1,2-ethyltoluene
b. 1,3-ethyltoluene
c. 1,4-ethyltoluene
d. 1,4-ethylmethylbenzene

18. What is the name of the aromatic compound formed in the following reaction?

$$+ \; CH_3CH_2Br \xrightarrow{AlBr_3}$$

a. Bromobenzene b. Toluene c. Ethylbenzene d. Ethyltoluene

19. What is the effect of delocalized electrons on the reactivity of aromatic compounds?
a. Delocalized electrons make aromatics more reactive than alkenes.
b. Delocalized electrons make aromatics less reactive than alkenes.
c. Delocalized electrons do not effect the reactivity of aromatic compounds.
d. Aromatic compounds will only undergo addition reactions because of delocalized electrons.

20. Toluene can be used as a gas additive to increase vehicle performance. If 500 mL of toluene is burned in your gas tank, how many grams of carbon dioxide will be produced? The density of toluene is approximately 0.50 g/mL.

a. 250 g CO_2 b. 119 g CO_2 c. 308 g CO_2 d. 836 g CO_2

Answers to Sample Test:

1. c
2. d
3. c
4. b
5. c
6. d
7. a

8. b
9. a
10. d
11. b
12. c
13. a
14. b
15. b
16. d
17. c
18. c
19. b
20. d

Chapter 10: Alcohols, Ethers, Thiols, Amines, and Organohalogens

Chapter Objectives:

1. To recognize and name organic molecules containing a hydroxyl functional group.
2. To become familiar with organic compounds containing oxygen bound to two carbon atoms.
3. To become familiar with organic compounds containing the sulfhydryl functional group.
4. To identify, name, and react organic compounds containing nitrogen.
5. To distinguish between alkyl halides and aryl halides.

Key Terms: The key terms are listed in the order in which they are encountered in the chapter.

Hetero-atom. An atom other than carbon or hydrogen that is commonly found as part of a functional group in an organic molecule.

Organohalogens. A molecule that contains a halogen atom bonded to an alkyl group or aromatic ring.

Hydroxyl group. The hydroxyl functional group consists of an oxygen atom bound to hydrogen (-OH).

Alcohol. A molecule that contains a hydroxyl group attached to a carbon atom of an alkyl group.

Phenol. An aromatic hydrocarbon containing a hydroxyl group attached to a carbon atom of the ring.

Glycols. Alcohols containing two hydroxyl groups.

Primary alcohol. A molecule that contains a hydroxyl group bonded to a carbon atom that is bonded to one other carbon atom with the exception of methanol that is also a primary alcohol.

Secondary alcohol. A molecule that contains a hydroxyl group bonded to a carbon atom that is directly attached to two other carbon atoms.

Tertiary alcohol. A molecule that contains a hydroxyl group bonded to a carbon atom that is directly attached to three other carbon atoms.

Disinfectant. A substance used to kill bacteria.

Dehydration. The loss of a water molecule that occurs when a hydroxyl group of an alcohol is eliminated along with a hydrogen atom on an adjacent carbon atom to produce an alkene.

Alcohol oxidation. The loss of a hydrogen atom from the hydroxyl group of an alcohol and a hydrogen atom from the hydroxyl bearing carbon atom to produce an aldehyde or ketone and a molecule of hydrogen.

Acid-base reaction of phenol. Phenol will lose a hydrogen atom in the presence of strong base to produce a salt and water.

Esterification. The reaction of an alcohol with a carboxylic acid to produce an ester.

Ether. An organic compound that contains an oxygen atom bonded to two carbon atoms.

Local anesthetic. A compound that is used to reduce or eliminate pain where it is administered.

General anesthetic. A compound that causes loss of sensation and loss of consciousness.

Sulfhydryl group. The –SH functional group.

Thiol. An organic compound containing a sulfhydryl functional group.

Mercaptan. An old name for a thiol.

Disulfide. The general term for a compound formed from the oxidation of two thiols.

Disulfide bond. The covalent linkage between two sulfur atoms joined by the oxidation of two thiols.

Amines. Organic compounds that contain a nitrogen atom bonded to a carbon atom.

Amino group. The functional group of amines that contains a nitrogen atom bonded to two hydrogen atoms (-NH_2).

Primary amine. An amine with two hydrogen atoms and one other group attached.

Secondary amine. An amine with one hydrogen atom and two other groups attached.

Tertiary amine. An amine with three groups attached.

Quaternary ammonium salt. A nitrogen atom with four groups attached and a positive charge that is counter-balanced by an anion.

Alkaloids. Nitrogen containing (nitrogenous) bases found in plants.

Heterocyclic ring. A mono- or polycyclic ring system that contains a hetero-atom in the ring.

Aniline. The simplest of aromatic amines with an amino group bonded to an aromatic ring.

Ammonium. The name given to a nitrogen atom containing four bonds to hydrogen and a positive charge.

Amides. Organic compounds that contain a nitrogen atom adjacent to a carbonyl group.

Alkyl halide. A halogen-containing organic compound that has a halogen directly attached to an alkyl group (R-X).

Aryl halide. An aromatic compound with a halogen atom directly attached to the ring.

Fluoro-. The prefix used to signify that a fluorine atom is present in a molecule.

Chloro-. The prefix used to signify that a chlorine atom is present in a molecule.

Bromo-. The prefix used to signify that a bromine atom is present in a molecule.

Iodo-. The prefix used to signify that an iodine atom is present in a molecule.

Synthesis. The building of organic molecules.

Chapter 10 Objective Details:

a. Alcohols and phenols. Alcohols are hydrocarbons that contain a hydroxyl functional group and phenols are aromatic rings that contain a hydroxyl functional group. The nomenclature for alcohols is the same as that for alkanes with minor exception. The alkane name is used with the –e at the end of the name being replaced by the suffix –ol. If the compound contains more than one hydroxyl group, the suffixes –diol and –triol are used and the suffix –e is retained. Phenol is the name given to the simplest of aromatic alcohols. Derivatives of phenol are named assuming that the hydroxyl group is in the one position on the ring. The prefixes ortho-, meta-, and para- are still used in the case of disubstituted aromatics. Alcohols can be primary (1°), secondary (2°), or tertiary (3°) depending on whether the hydroxyl group is bonded to a carbon atom that is attached to one, two, or three other carbon atoms, respectively.

The properties of alcohols are different from hydrocarbons in that the hydroxyl functional group is polar allowing for the presence of hydrogen bonding and dipole-dipole interactions in polar solvents. These intermolecular interactions are stronger than the London dispersive forces that hold together alkanes and alkenes making alcohols of comparable size boil at higher temperatures. The larger the hydrocarbon chain in an alcohol, the lower the solubility in polar solvents. The hydrocarbon chain is nonpolar and will eventually override the effect of the polar hydroxyl group for alcohols containing more than three carbon atoms.

Alcohols are susceptible to dehydration, oxidation, combustion, and esterification. Phenol is slightly acidic due to the delocalized electrons of the aromatic ring. As an acid, phenol will react with a strong base to produce a salt and water. The dehydration of an alcohol is an acid catalyzed reaction that results in the formation of an alkene. Water will be lost from the elimination of the hydroxyl group from an alcohol and a hydrogen atom on an adjacent carbon atom that contains the fewest hydrogen atoms. This is sometimes referred to as the anti-Markovnikov rule or Saytzeff's rule. The oxidation of an alcohol involves the loss of a hydrogen atom from the hydroxyl group and a hydrogen atom from the hydroxyl bearing carbon atom. The result of alcohol oxidation is molecular hydrogen and either an aldehyde or ketone. Combustion reactions of alcohols involve the addition of oxygen to produce carbon dioxide, water, and energy. Esterification reactions of alcohols involve the reaction of the alcohol with a carboxylic acid to produce an ester. A summary of key reactions is provided for you. Certain atoms are circled to indicate that they are the ones involved in the reaction.

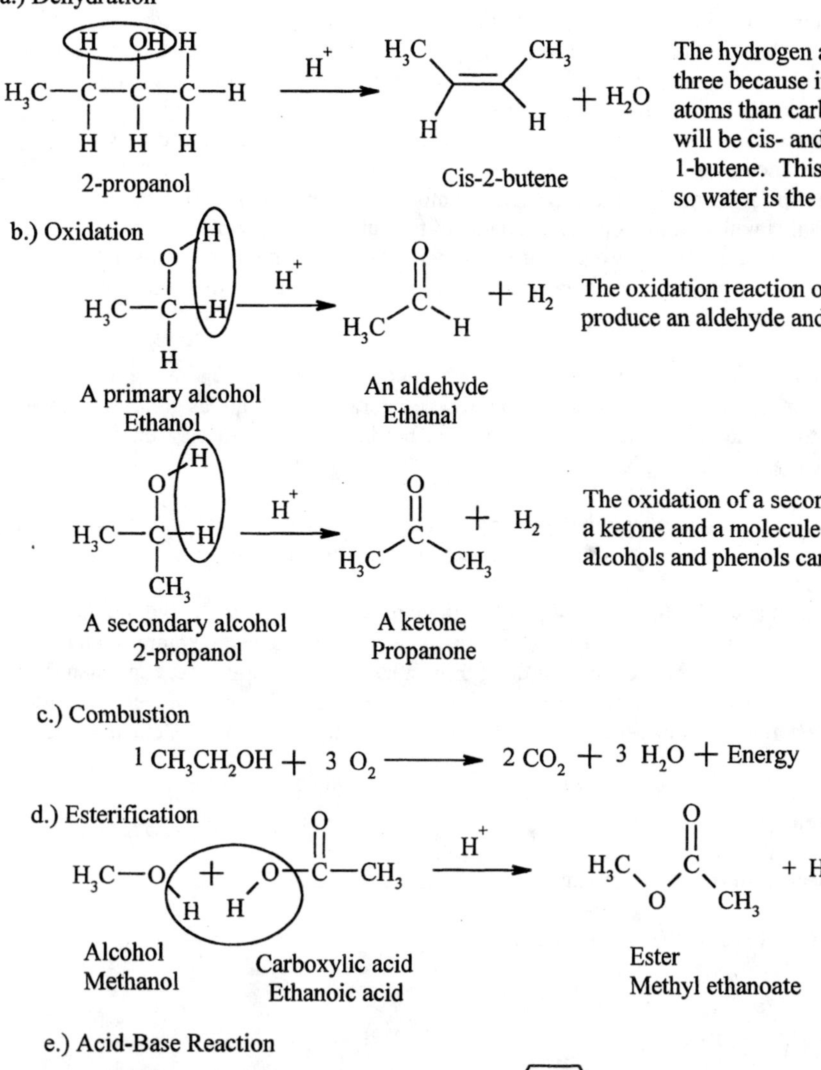

The hydrogen atom is taken from carbon three because it contains fewer hydrogen atoms than carbon one. The main products will be cis- and trans-2-butene rather than 1-butene. This is a dehydration reaction, so water is the other product.

The oxidation reaction of a primary alcohol will produce an aldehyde and a molecule of hydrogen.

The oxidation of a secondary alcohol will produce a ketone and a molecule of hydrogen. Tertiary alcohols and phenols cannot be oxidized.

$$1\ CH_3CH_2OH + 3\ O_2 \longrightarrow 2\ CO_2 + 3\ H_2O + \text{Energy}$$

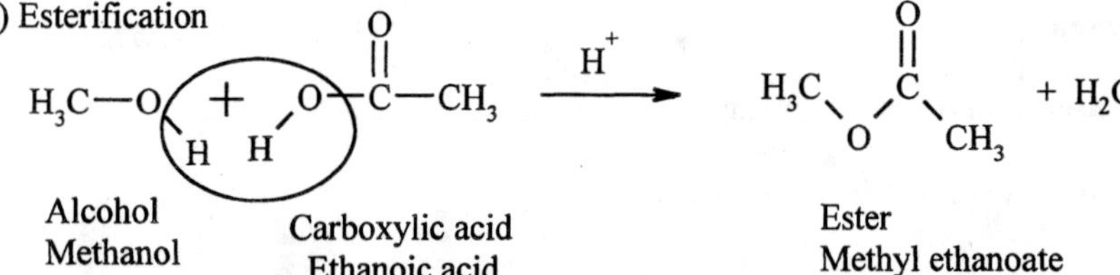

e.) Acid-Base Reaction

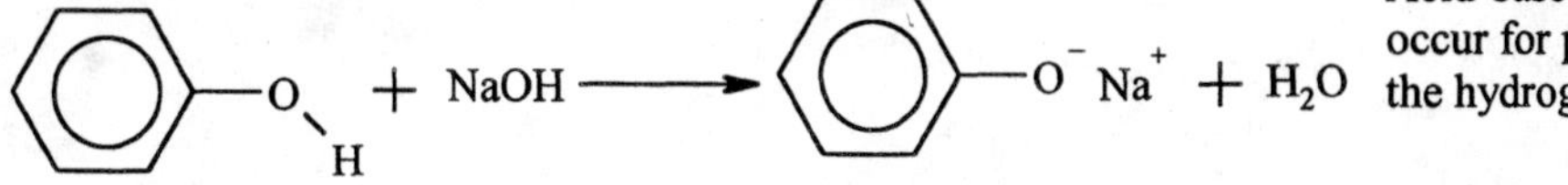

Acid-base reactions only occur for phenols because the hydrogen atom is acidic.

b. Ethers. Ethers are compounds that contain an oxygen atom bound to two carbon atoms(R-O-R'). Ethers are typically named by an old system of nomenclature that lists the alkyl or aromatic groups attached to either side of the oxygen atom in alphabetical order and puts the word ether at the end. The oxygen atom of an ether is surrounded by nonpolar hydrocarbons that make only small ethers water-soluble. There are enough dipole-dipole and hydrogen bonding interactions in the small ethers to be soluble in water, but as the alkyl groups get bigger, the ethers behave like alkanes. Ethers are primarily known for their anesthetic properties and are often used as fuel additives for their ability to burn well.

c. Thiols. Thiols are compounds containing a sulfhydryl functional group (-SH). The nomenclature for thiols is the same as for alcohols except the suffix –e is retained and the –ol of the alcohol is replaced with –thiol. Sulfur is less electronegative than oxygen, so thiols are not as water-soluble nor do they boil as at high a temperatures as alcohols of comparable size. Thiols can be oxidized to form disulfides and disulfide bonds can be broken by reduction.

d. Amines. Amines are organic compounds that contain a nitrogen atom bonded to at least one carbon atom. The functional group of amines is termed the amino group ($-NH_2$). Like alcohols, amines can exist as primary, secondary, or tertiary to indicate that one, two, or three carbon atoms are directly bonded to the nitrogen atom, respectively. Unlike alcohols, the amino nitrogen can form a quaternary ammonium salt. These salts are formed when the nitrogen atom of an amine binds to four groups. Amines can be named in two different ways. The IUPAC method includes the full name of the parent hydrocarbon chain with the amine considered as an amino group for primary amines. If there are more than one alkyl group is attached to the nitrogen atom, then a capital N with a dash precedes the name of the substituent to indicate that the substituent is bonded directly to the nitrogen atom. The other method calls for the substituents to be listed in alphabetical order and the word amine added at the end. In the case of aromatic amines, the parent structure is called aniline.

The physical properties of primary amines are similar to those of alcohols, but because nitrogen is not as electronegative, the effects of hydrogen bonding and dipole-dipole interactions are not as great. The lower electronegativity of the nitrogen atom is responsible for the lower boiling points and decreased water-solubility at longer hydrocarbon chain lengths compared to alcohols.

Amines are by nature alkaline molecules that will deprotonate an acid to form a quaternary ammonium salt. The nomenclature for these salts is to use replace the suffix –amine with –ammonium.

e. Organohalogens-alkyl halides and aryl halides. You are, by this time, already familiar with alkyl halides and aryl halides. These are compounds that contain a halogen directly attached to the carbon atom of an alkyl group or aromatic ring. The IUPAC nomenclature for these molecules lists the halogen first and then the parent molecule. The common names for these compounds list the alkyl group first and the halogen ending in –ide second. The prefixes fluoro-, chloro-, bromo-, and iodo- are used to represent the elements F, Cl, Br, and I, respectively.

Solutions to Odd Numbered Problems:

1. The hydrocarbons are the only class of organic compound that does not contain a hetero-atom.

3. Phenols are different than alcohols because they have an aromatic ring.

C_6H_5-OH $\qquad$ H_3C-OH

Aromatic ring
Phenol

No aromatic ring
Alcohol

5.

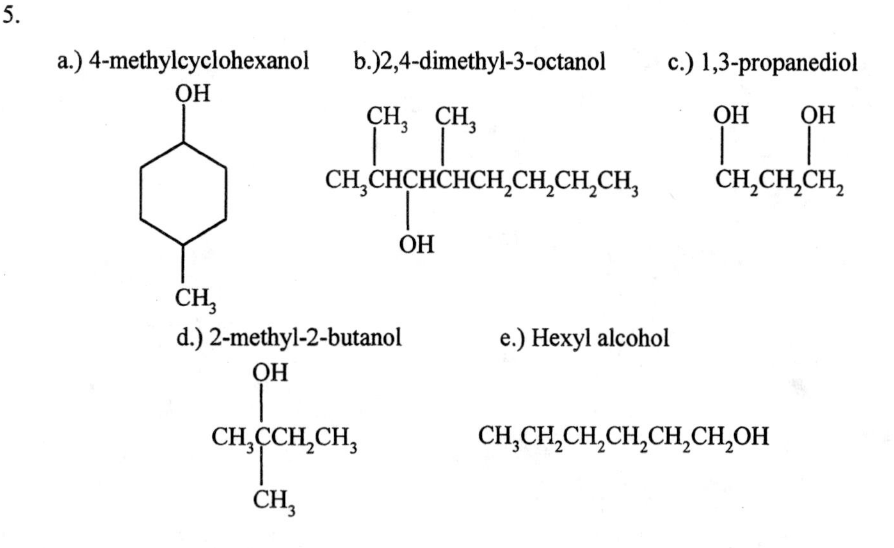

7. a. 2,4,4-trimethyl-2-pentanol
   b. 3-methyl-2-butanol
   c. 2,2,5-trimethylcyclohexanol
   d. 1,6-hexanediol
   e. 3-ethyl-3-pentanol

9. a. Primary
   b. Secondary
   c. Tertiary
   d. Primary

11. a. 4-chlorophenol or p-chlorophenol
    b. 2-ethylphenol or o-ethylphenol
    c. 2,4-dichlorophenol

13.

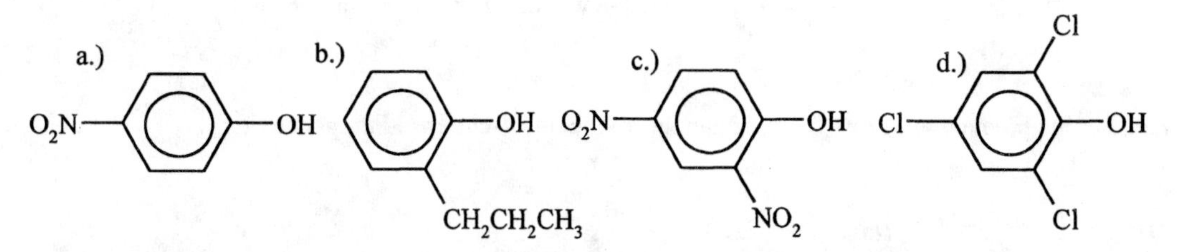

15. Both methanol and ethanol form hydrogen bonding interactions and dipole-dipole interactions with water. Methane and ethane are both nonpolar and do not have any way to interact with a polar solvent.

17. Phenol and small phenols are water soluble, but as the size of the nonpolar hydrocarbons attached to the aromatic ring increases, water solubility decreases. The more nonpolar atoms a molecule has, the more difficult it is for that molecule to be soluble in a polar solvent.

19. Carbolic acid was the original disinfectant. Carbolic acid is another name for what is now known as phenol. Phenol is no longer used as a disinfectant because it is a skin irritant and smells bad.

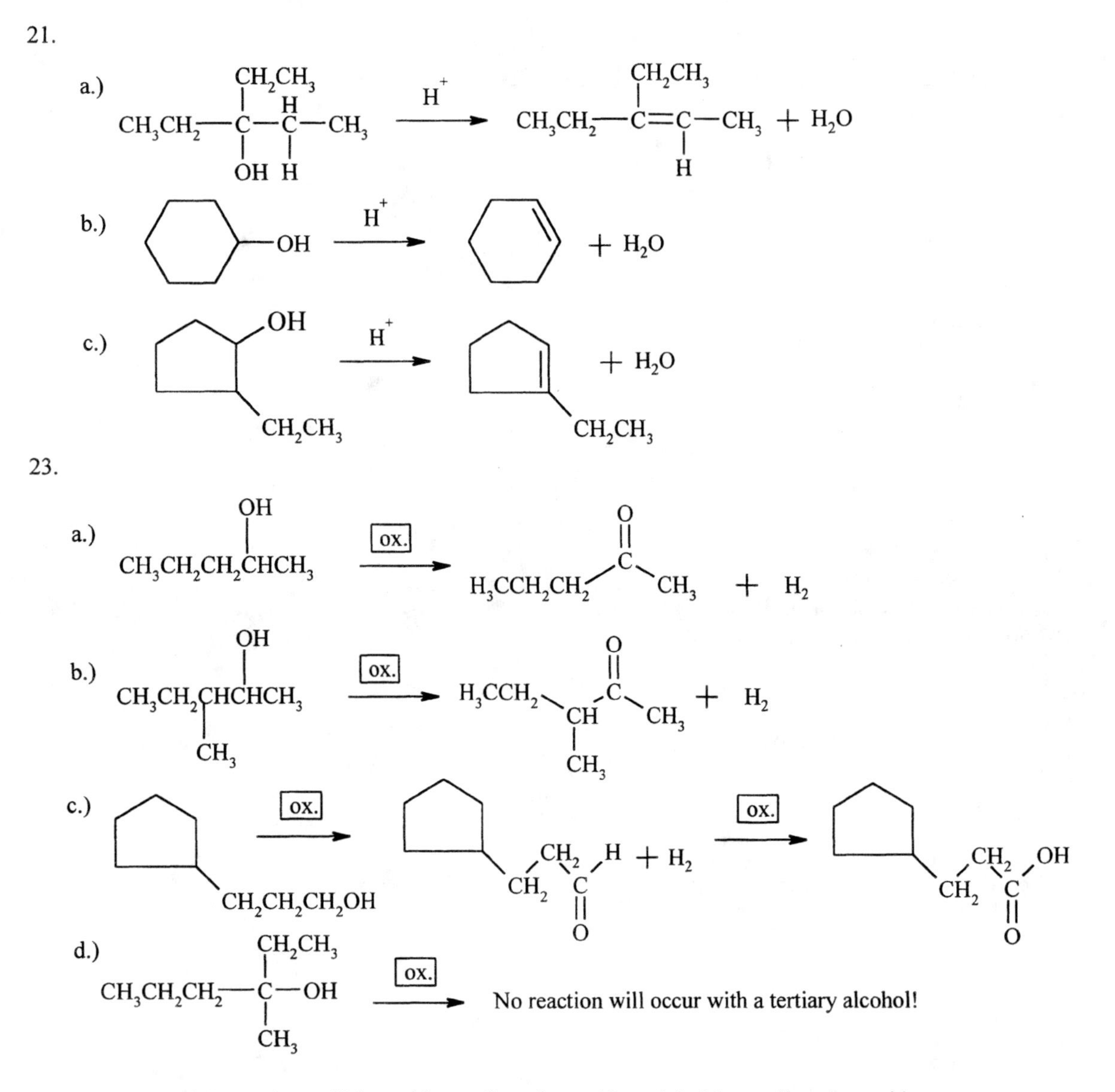

25. The name of the product will be p-chlorosodiumphenoxide or 1,4-chlorosodiumphenoxide.

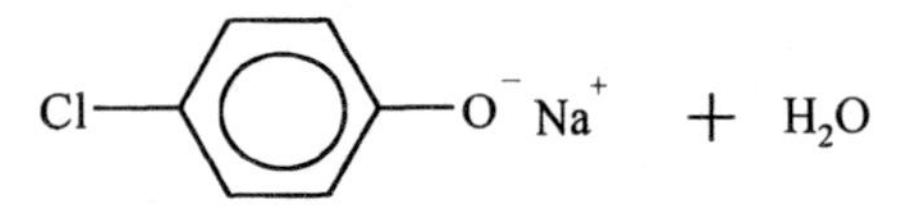

27. The functional group that is associated with ethers is an oxygen atom.

29. a. Butyl methyl ether
    b. Ethyl propyl ether
    c. Cyclopentyl ethyl ether
    d. Dipropyl ether

31.

a.) Isopropyl phenyl ether b.) Cyclohexyl propyl ether c.) Diphenyl ether

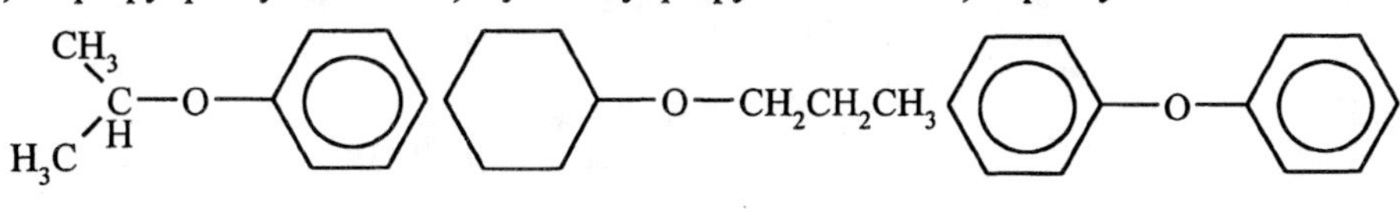

d.) Butyl isopropyl ether

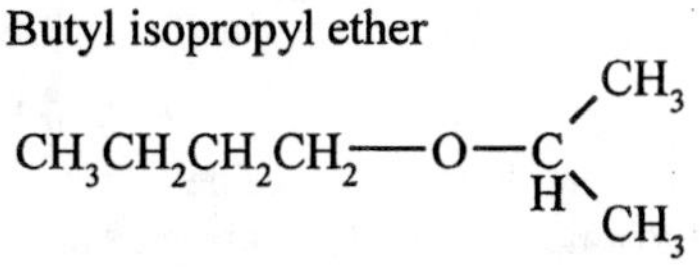

$$CH_3CH_2CH_2CH_2-O-\underset{H}{C}(CH_3)_2$$

33. The oxygen atom in an ether is able to hydrogen bond to a limited extent with water molecules. There is also a dipole moment that is created due to the electronegativity of oxygen that contributes to dipole-dipole interactions with water. These two intermolecular forces are not present in other nonpolar hydrocarbons and are what makes small ethers water-soluble.

35. a. The functional group of a thiol is the sulfhydryl group (-SH).
    b. Sulfur is not as electronegative as oxygen, so comparable size thiols will not be as water soluble nor as high in boiling point as alcohols.

37. a. Methanethiol
    b. 3-pentanethiol
    c. 4-ethyl-1-heptanethiol

39.

a.) $H_3C-\overset{H_2}{C}-C(CH_3)(SH)-\overset{H_2}{C}-CH_3$

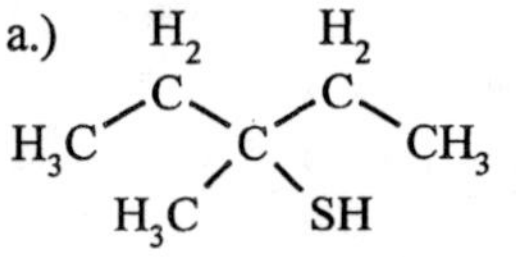

b.) $CH_3-CH(CH_3)-SH$ (CH-SH with two CH_3)

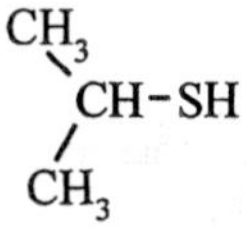

c.) HS—(benzene ring)

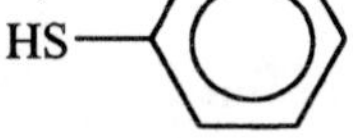

41.

$$H_3C-S-S-CH_3 \xrightarrow{\boxed{red.}} 2\ CH_3SH$$

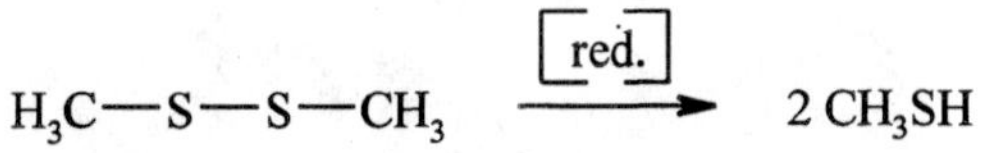

43. a. Primary
    b. Tertiary
    c. Secondary

45. A quaternary ammonium salt contains a nitrogen atom with four groups attached and a positive charge that is counter-balanced by the presence of an anion. Amines have nitrogen atoms with only three groups attached and no charge.

47. A hetercycle is a mono- or polycyclic molecule that contains a hetero-atom as part of the ring structure.

49. a. Propyl amine or aminopropane
    b. N,N-dicyclohexylamine
    c. N,N-diethylaniline

51. a. N-ethylaniline
    b. N,N-dimethyl-1-aminopentane
    c. 2-aminopropane

53.

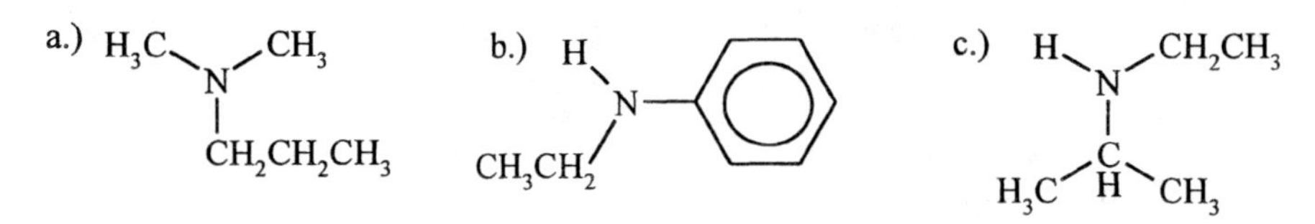

55. Hydrogen bonding

57. Organic ammonium salts contain a nitrogen atom bonded to four other groups. The formation of four bonds causes nitrogen to have a positive charge. Ammonium salts are formed from the reaction of an amine with an acid, so the portion of the acid that has been deprotonated will act as the anion for the ammonium salt. Most other salts are formed from the combination of a metal with a nonmetal.

59.

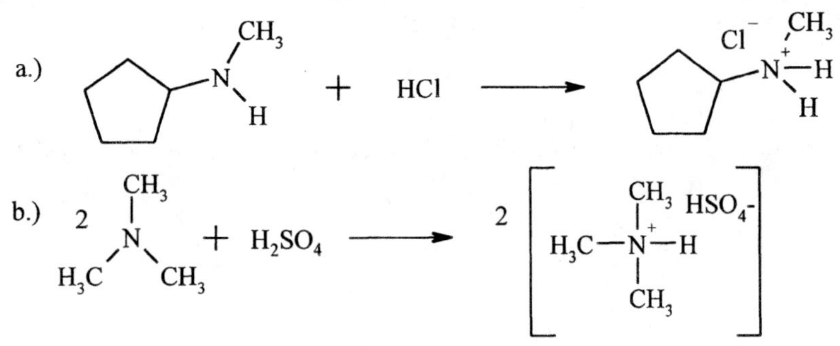

61. a. N-cyclohexyl-N-methylammonium chloride
    b. Trimethylammonium sulfate

63. An aryl halide contains an aromatic ring where an alkyl halide contains an alkyl group.

65. a. 1,2-dichloroethane
    b. Trichloromethane
    c. 1-bromo-2-methylcyclohexane

67.

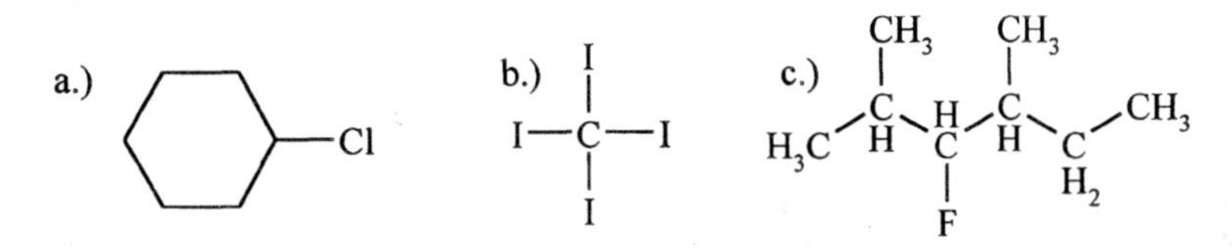

69.

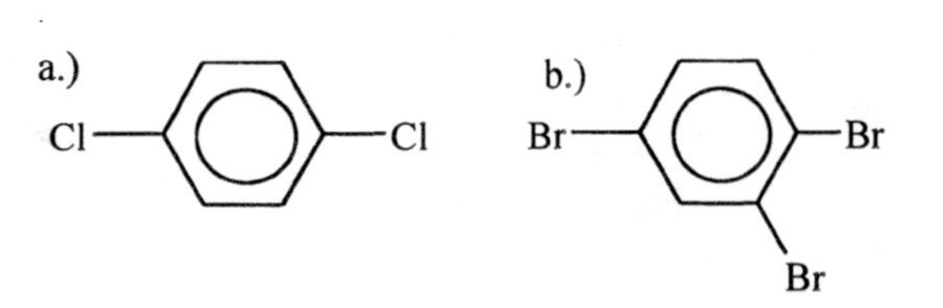

71. Organohalogens are inherently insoluble in polar solvents. The electronegative halogen atom is not enough to make the alkyl halide or aryl halide polar. Organohalogens thus make good nonpolar solvents.

73. Amino acids are called amino acids because they all contain both an amino group and a carboxyl group as part of their basic structure by definition. All amino acids contain an amino group, a carboxyl group, an alpha carbon, an alpha hydrogen and any one of twenty different groups attached to the alpha carbon atom. The hydroxyl-containing amino acids are serine and threonine. The amino acid that contains a thiol group is called cysteine.

Sample Test:

1. How can you tell the difference between a hydrocarbon and all other classes of organic compounds?
   a. The presence of hydrogen.
   b. The presence of carbon.
   c. The presence of a hetero-atom
   d. You can't!

2. Name the following alcohol.

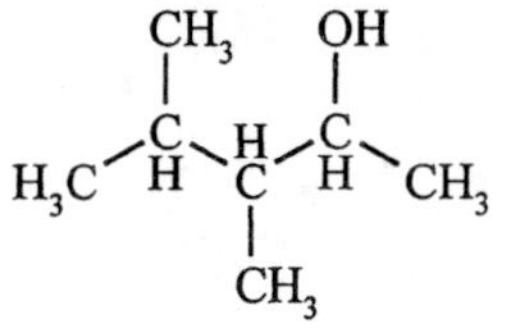

   CH_3 OH
   H_3C C H C H C H CH_3
   CH_3

   a. 2,3-dimethyl-4-pentanol
   b. 2,3-dimethyl-4-pentaneol
   c. 3,4-dimethyl-2-pentanol
   d. 3,4-dimethyl-2-pentaneol

3. What is the name for the following phenol?

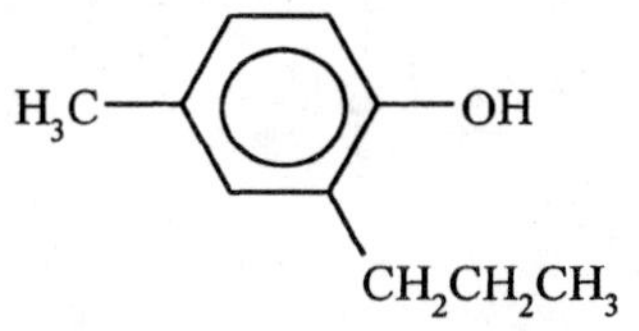

   H_3C OH
   $CH_2CH_2CH_3$

   a. 2-propyl-4-methylphenol
   b. 4-methyl-2-propylbenzene alcohol
   c. p-methyl-2-propylphenol
   d. 4-methyl-2-propylphenol

4. Which is the correct IUPAC name for 1-methyl-3-ethyl-4-cyclohexanol?
   a. 2-ethyl-4-methylcyclohexanol
   b. 4-methyl-2-ethylcyclohexanol
   c. o-ethyl-p-methylcyclohexanol
   d. 2-ethyl-4-methylphenol

5. What will be the major organic product in the dehydration of 2,3-dimethyl-3-pentanol?
   a. Trans-3,4-dimethyl-2-pentene
   b. Cis-3,4-dimethyl-2-pentene
   c. 2,3-dimethyl-2-pentene
   d. Both a and b are correct.

6. What kind of organic compound would originate from the oxidation of a secondary alcohol?
   a. Aldehyde b. Ketone c. Carboxylic acid d. Tertiary alcohol

7. Provide a name for the ether drawn below?

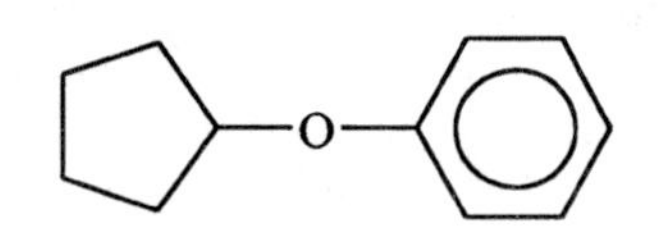

a. Cyclopentylphenol
b. Cyclopentyl benzene ether
c. Phenyl cyclopentyl ether
d. Cyclopentyl phenyl ether

8. What is the correct name for the following molecule?

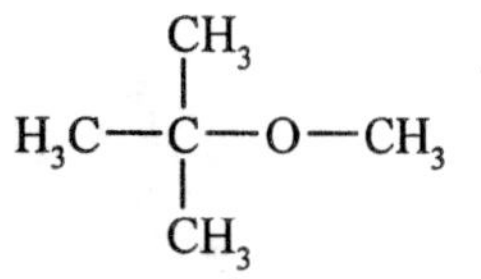

$$H_3C-\overset{\displaystyle CH_3}{\underset{\displaystyle CH_3}{\overset{|}{\underset{|}{C}}}}-O-CH_3$$

a. Methyl tert-butyl ether
b. 2,2-dimethyl propyl ether
c. isopropylmethyl methyl ether
d. isopropyl methyl ether

9. What is the name of the following organic molecule?

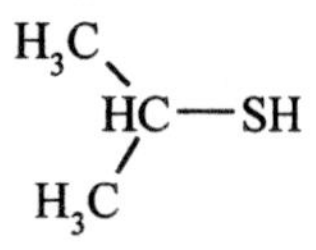

$$\begin{matrix} H_3C \\ \quad\diagdown \\ \quad HC-SH \\ \quad\diagup \\ H_3C \end{matrix}$$

a. Isopropylthiol b. 2-propanethiol c. Isopropyl mercaptan d. More than one correct answer.

10. What were the reactants and conditions that gave the following product?

$$H_3CCH_2-S-S-CH_2CH_3$$

a. Reduction of two molecules of ethanethiol.
b. Oxidation of two molecules of ethanethiol.
c. Reduction of two molecules of methanethiol.
d. Oxidation of two molecules of methanethiol.

11. What type of compound is N,N-diethylaniline?

a. Primary amine b. Secondary amine c. Tertiary amine d. Quaternary ammonium salt

12. What is the name of the following compound?

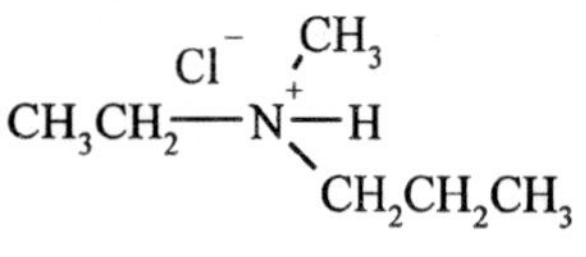

$$Cl^- \quad CH_3CH_2-\overset{+}{N}(CH_3)(CH_2CH_2CH_3)-H$$

a. Ethylmethylpropylamine
b. Chloroethylmethylpropylamine
c. Ethylmethylpropylammonium
d. Ethylmethylpropylammonium chloride

13. Which of the following compounds would have the lowest boiling point?
    a. Methane b. Methanol c. Dimethyl ether d. Methylamine

14. What is the correct structure of m-bromoanisol?

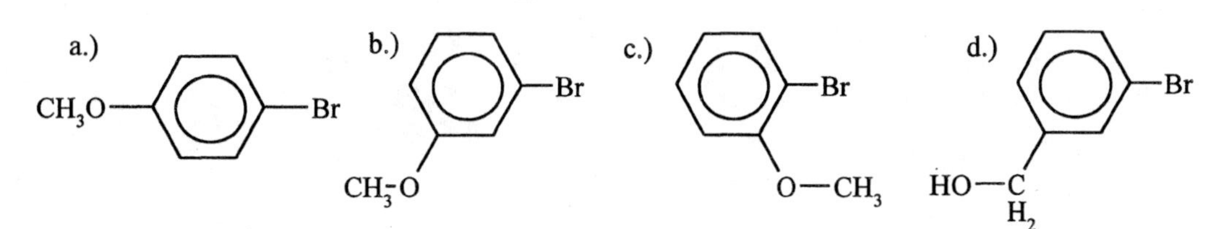

15. What is the name of the following alkyl halide?

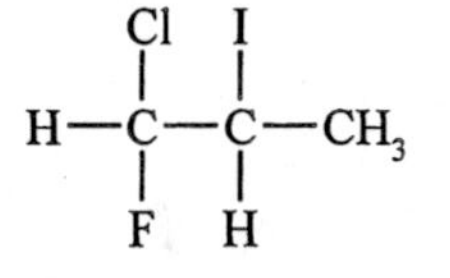

a. Fluorochloroiodopropane
b. Chlorofluoroiodopropane
c. 1-iodo-2-fluoro-2-chloropropane
d. 1-chloro-1-fluoro-2-iodopropane

16. What type of alcohol is shown below?

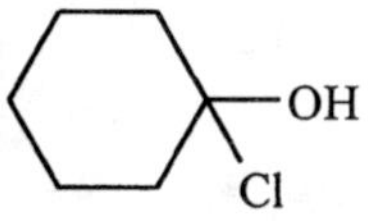

OH
Cl

a. Primary b. Seconary c. Tertiary d. Quaternary

17. Name the following molecule.

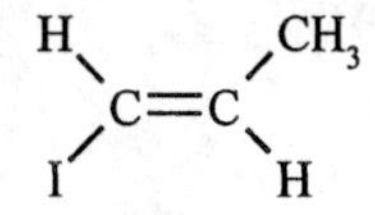

a. Iodopropene
b. Cis-1-iodo-1-propene
c. 1-iodo-1-propene
d. Trans-1-iodo-1-propene

18. How could you make a molecule of 2-butene?
    a. Oxidation of butanol.
    b. Oxidation of 2-butanol
    c. Dehydration of butanol
    d. Dehydration of 2-butanol

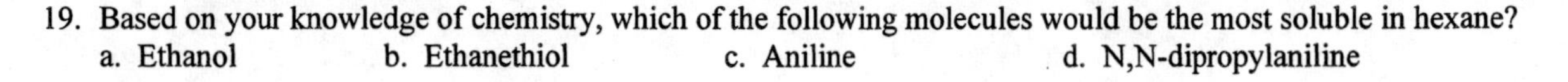

19. Based on your knowledge of chemistry, which of the following molecules would be the most soluble in hexane?
    a. Ethanol b. Ethanethiol c. Aniline d. N,N-dipropylaniline

20. Ether is an anesthetic, but it is no longer used partly because it is so flammable. How many moles of oxygen will react in the complete combustion of 1.00 mole of diethyl ether?

    a. 13.0 mol O_2 b. 6.00 mol O_2 c. 2.00 mol O_2 d. 1.00 mol O_2

Answers to Sample Test:

1. c
2. c
3. d
4. a
5. c
6. b
7. d
8. a
9. d
10. b
11. c
12. d
13. a
14. b
15. d
16. b
17. d
18. d
19. d
20. b

Chapter 11: Aldehydes, Ketones, Carboxylic Acids, Esters, and Amides

Chapter Objectives:

1. To distinguish between compounds containing carbon-oxygen double bonds.
2. To name, draw, and react simple aldehydes.
3. To name, draw, and react ketones.
4. To identify and understand the reactivity of carboxylic acids.
5. To distinguish and name carboxylic acid derivatives.

Key Terms: The key terms are listed in the order in which they are encountered in the chapter.

Carbonyl. The carbon-oxygen double bond that is the functional group or part of the functional group for many classes of organic compounds.

Aldehydes. The class of organic compound that has the carbonyl carbon atom bonded to a hydrogen atom on one side and a hydrogen atom or a carbon atom on the other side.

Ketone. The class of organic compound that has the carbonyl carbon atom bonded to two carbon atoms.

Carboxylic acid. The class of organic compound that has the carbonyl carbon atom bonded to a hydroxyl group on one side and a hydrogen atom or carbon atom on the other side.

Ester. The class of organic compound that has the carbonyl carbon atom bonded to an oxygen atom that is attached to a carbon atom.

Amides. The class of organic compound that has the carbonyl carbon atom bonded to a nitrogen atom.

Hydroxy. The term given to the hydroxyl group of a compound when it is a substituent.

Benzaldehyde. The simplest aromatic aldehyde that also serves as the parent name for all aromatic aldehydes.

Aldehyde reduction. The reduction of an aldehyde adds a molecule of hydrogen across the carbonyl to produce a primary alcohol.

Aldehyde oxidation. The oxidation of an aldehyde adds an oxygen atom to produce a carboxylic acid.

Tollen's test. Silver ions are used to oxidize aldehydes to produce carboxylic acids and a coating of shiny silver metal.

Reducing sugar. A reducing sugar is one that contains an aldehyde group that can be oxidized.

Hemiacetal. The product of a reaction between an aldehyde or ketone and an alcohol that contains a carbon atom single bonded to two oxygen atoms in place of the carbon-oxygen double bond of the reacting carbonyl compound.

Acetal. The product of the reaction between a hemiacetal and an alcohol where the hydroxyl group of the hemiacetal is converted to an alkoxy group (oxygen bonded to an alkyl group).

Acetophenone. The simplest of aromatic ketones (methyl phenyl ketone) that is the root for other aromatic ketones.

Carboxyl group. The functional group of a carboxylic acid that contains a carbonyl group that has a hydroxyl group attached to it.

Benzoic acid. An aromatic ring with a carboxyl group attached that serves as the parent structure for aromatic acid derivatives.

Carboxylate anion. The anion formed when a carboxylic acid donates a hydrogen ion.

Acidosis. The abnormal lowering of blood pH as a result of the too much acid offsetting the buffering capacity of blood.

Lactic acidosis. The condition that develops when lactic acid exceeds normal ranges due to excessive exercise or hyperventilation.

Diabetic acidosis. The condition that develops when ketone bodies exceed normal ranges due to a diabetic crisis or extreme fasting.

Esterification. The acid-catalyzed reaction of a carboxylic acid with an alcohol to form an ester and water.

Nylon. A polymer formed by the reaction of a diamine with a diacid chloride.

Hydrolysis. The acid or base catalyzed cleavage of an ester by the addition of a water molecule to yield a carboxylic acid and an alcohol.

Saponification. The base catalyzed hydrolysis of an ester to produce an alcohol and a carboxylate ion.

Analgesic. A substance that relieves pain.

Amide. The class of organic compounds formed from the reaction of a carboxylic acid with either ammonia or an amine.

Acid chloride. A compound containing a chlorine atom attached to a carbonyl group.

Anhydrous. A compound that does not contain water.

Acid anhydride. The type of substance that forms when two carboxylic acids react and lose water.

Thiol ester. The product of the reaction between a carboxylic acid and a thiol.

Chapter 11 Objective Details:

a. The carbonyl group. The carbonyl group consists of a carbon atom double bonded to an oxygen atom. Among the carbonyl-containing compounds are the aldehydes, ketones, carboxylic acids, esters, and amides. These compounds differ only in the attachment of atoms to the carbonyl carbon atom. The simplest of carbonyl-containing compounds are the aldehydes that have a hydrogen atom attached to the carbonyl carbon atom. Ketones have a carbon atom attached on either side of the carbonyl carbon atom. Carboxylic acids have a hydroxyl group bonded to the carbonyl carbon atom. Esters have a oxygen atom bonded to the carbonyl carbon atom on one side an a carbon atom on the other side. Amides have a nitrogen atom bonded to the carbonyl carbon atom.

b. Aldehydes. Aldehydes are carbonyl containing compounds that have a hydrogen atom directly bonded to the carbonyl carbon atom. This means that the carbonyl carbon atom is always at one end of the parent chain. The carbonyl carbon atom is the first atom counted as part of the longest continuous chain of carbon atoms. Aldehydes are named in the same way as alkanes, except the suffix –e is replaced with the suffix –al. The common names for aldehydes are formaldehyde, acetaldehyde, propionaldehyde, and butyraldehyde to represent aldehydes containing 1, 2, 3, and 4 carbon atoms, respectively. The IUPAC names for these first four aldehydes would be methanal, ethanal, propanal, and butanal.

Aldehydes are slightly polar due to the difference in electronegativity between carbon and oxygen. As the size of the alkyl group or aryl group attached to the carbonyl carbon atom increases, the solubility of these compounds in polar solvents decreases. Aldehydes have lower boiling points than alcohols of equivalen size because they cannot hydrogen bond to other aldehydes where alcohols can hydrogen bond to other alcohols.

Aldehydes can be reduced to form primary alcohols or oxidized to form carboxylic acids. The facile oxidation reaction makes aldehydes easy to detect in the presence of the metal ions Ag^+ (Tollen's test) or Cu^{2+} (Benedict's and Fehling's tests). When aldehydes react with alcohols, they form hemiacetals that can react with another alcohol to form an acetal.

c. **Ketones.** Ketones are organic compounds that have two carbon atoms bonded to the carbonyl carbon atom. They are named as alkanes with the suffix –e replaced by the suffix –one. The common names of ketones include the alkyl groups on either side of the carbonyl carbon atom being listed in alphabetical order followed by the word ketone. The simplest of ketones is ethanone also commonly called acetone. Aromatic ketones are typically derivatives of acetophenone that is not often referred to as methyl phenyl ketone. Benzophenone is the ketone containing two aromatic rings bonded to the carbonyl group. Ketones are only slightly polar due to the difference in electronegativity of the carbon and oxygen atoms of the carbonyl group. Ketones can be reduced to secondary alcohols and oxidized by combustion. The reaction of a ketone with an alcohol will yield a hemiacetal that can be converted to an alcohol by further reaction with an alcohol.

d. **Carboxylic acids.** Carboxylic acids are weak organic acids that contain the carboxy functional group (-COOH). Upon deprotonation of the acid, a carboxylate ion is formed ($-COO^-$). Carboxylic acids are named as the parent alkane dropping the suffix –e and adding the suffix –oic acid. Carboxylate ions are named as the parent alkane dropping the suffix –e and adding the suffix –oate. The common names for carboxylic acids containing one, two, three, and four carbon atoms have the prefixes form-, acet-, propion-, and butyr-, respectively. In the common names, substituents are frequently listed using the Greek alphabet rather than the IUPAC numbering system. The letters in the Greek alphabet are used to represent substituents in the 2, 3, 4, 5, etc. positions as α, β, δ, γ, etc., respectively. It is assumed that the carbon atom of the carboxylate group is always in the first or one position. Aromatic carboxylic acids are derivatives of benzoic acid.

Carboxylic acids are even more polar than alcohols of comparable size. The carboxyl group is able to hydrogen bond making the boiling points for these acids relatively high. The solubility of carboxylic acids in aqueous solution is good up to hydrocarbon chain lengths of 5 carbon atoms.

The reaction of a carboxylic acid and a base will produce a carboxylate ion. The nomenclature for these types of compounds is the same as that of salts where the cation is named prior to the anion. The acid catalyzed reaction of a carboxylic acid with an alcohol will produce an ester and water. The reaction of an acid chloride and an amine will produce an amide and a hydrohalogen.

e. **Esters, amides, and other derivatives of carboxylic acids.** Esters are carbonyl containing organic compounds formed from the reaction of a carboxylic acid with an alcohol. The name of the ester begins with the name of the alcohol minus the suffix –ol and with the suffix –yl. The name of the carboxylic acid follows the name of the alcohol with the suffix –oic acid replaced by the suffix –oate. Esters can undergo hydrolysis reactions that are just as the word says. A hydrolysis reaction is a "hydro" (for water) "lysis" (for breaking apart) reaction. The base catalyzed hydrolysis of an ester is known as a saponification reaction.

Amides are compounds formed from the reaction of a carboxylic acid and either ammonia (NH_3) or an amine. Amides are named by using the name of the parent carboxylic acid and replacing the suffix –oic acid with the suffix –amide. If the amine that makes the amide is primary or secondary, the substituents attached to the nitrogen atom have a capital N separated from the name of the substituent by a hyphen. Amides can be hydrolyzed to their parent acid and amine in acidic aqueous conditions.

Acid chlorides, acid anhydrides, and thiol esters are all derivatives of carboxylic acids. Acid chlorides are the result of a carboxylic acid containing a chlorine atom in place of the hydroxyl group. Acid anhydrides are compounds formed from the removal of water from the reaction of two carboxylic acids. Thiol esters are formed from the reaction of a carboxylic acid with a thiol.

Solutions to Odd Numbered Problems:

1. A carbonyl group consists of a carbon atom double bonded to an oxygen atom. The structure is C=O.

3. An aldehyde will always have a hydrogen atom directly attached to the carbonyl carbon atom. A ketone will always have the carbonyl carbon atom bonded to two other carbon atoms.

5. a. 3,5-dimethylheptanal
   b. 2,3-dichloropentanal
   c. 2,3-dibromobenzaldehyde

7. a. 2-bromopropanal or 2-bromopropionaldehyde
   b. 4-hydroxybutanal or 4-hydroxybutyraldehyde
   c. 4-chlorobenzaldehyde or p-chlorobenzaldehyde

9.

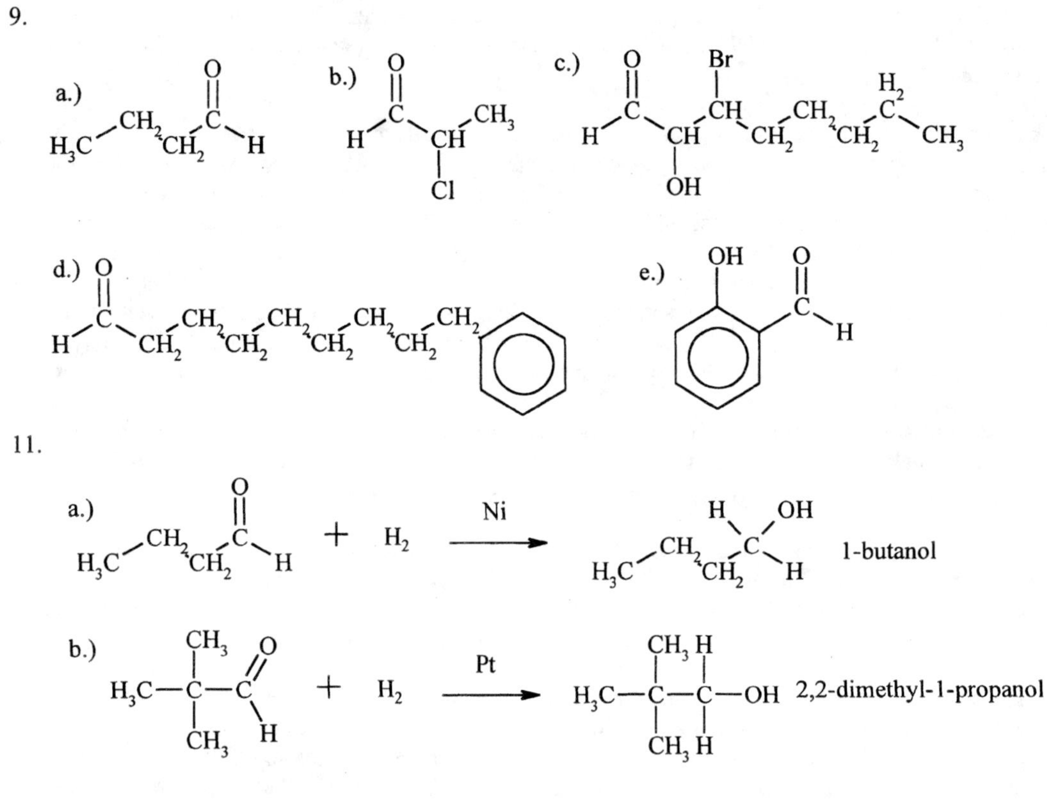

13. $CH_3(CH_2)_7CHO$ or nonanal

15.

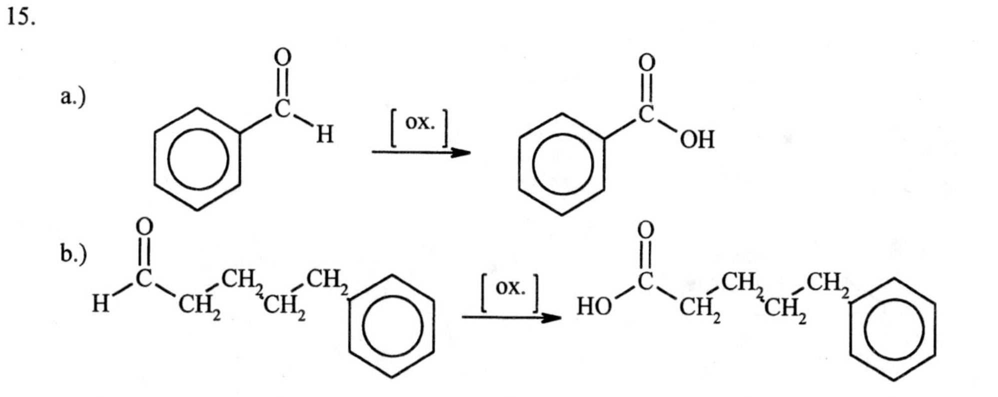

17. a. Positive
    b. Negative
    c. Negative
    d. Positive

19. Hemiacetals are produced by the reaction of an aldehyde and an alcohol.

21.

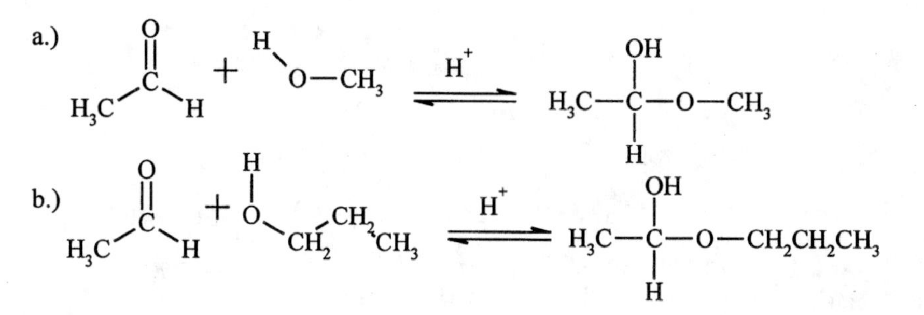

23. a. 2-pentanone
    b. 3,5-dimethyl-2-hexanone
    c. Acetophenone
    d. Chloropropanone

25. a. Methyl propyl ketone
    b. Diethyl ketone

27.

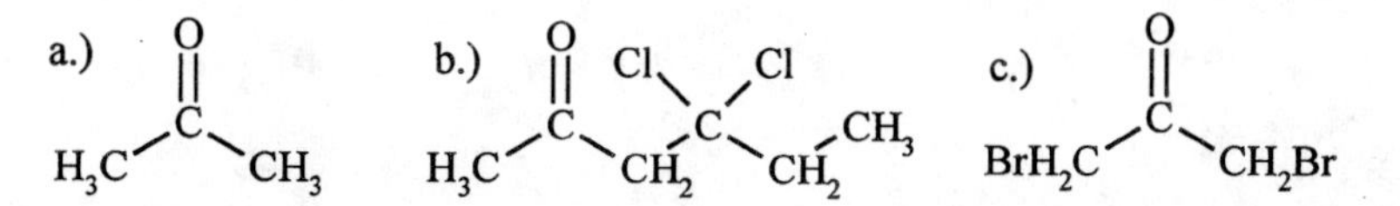

29. The correct name should be 4-octanone. The carbonyl group should be given the lowest number possible.

31. The reduction of a ketone will result in the formation or a secondary alcohol.

33.

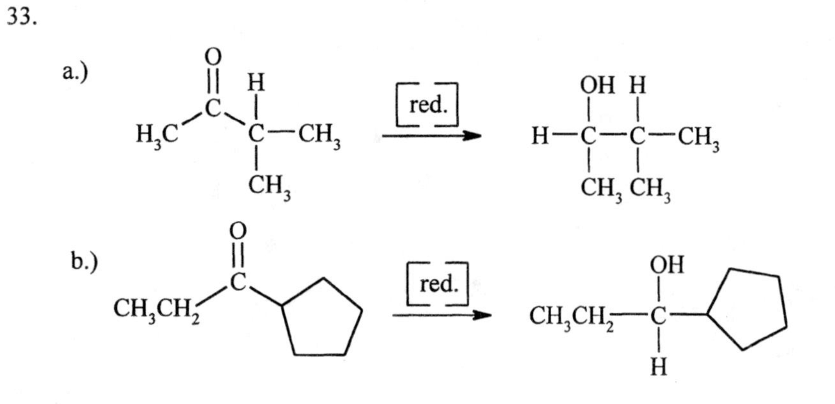

35. Propanone

37. Propanone is the ketone and ethanol is the alcohol.

39. The hydrogen atom on the hydroxyl group of the carboxyl group is donated.

41. a. 3-hydroxypentanoic acid
    b. 3-bromobenzoic acid
    c. 2,2-dimethylpropanoic acid
    d. 2-ethylbutanoic acid

43.

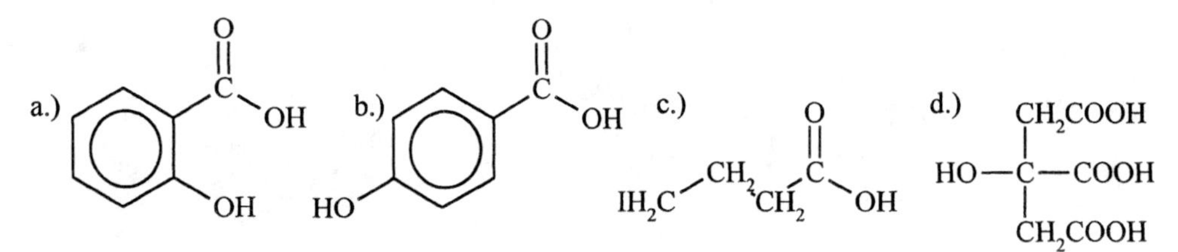

45. a. Ethanoate or acetate
    b. Phenyl methanate or benzoate
    c. Butanoate or butyrate

47. a. Sodium ethanoate or sodium acetate
    b. Sodium phenyl methanoate or sodium benzoate
    c. Sodium butanoate or sodium butyrate

49.

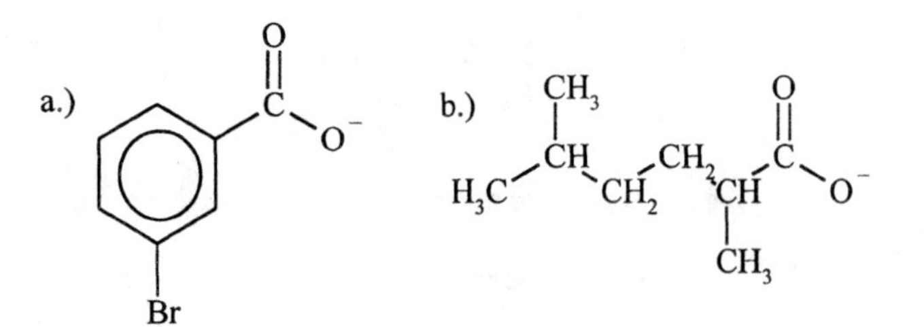

51.

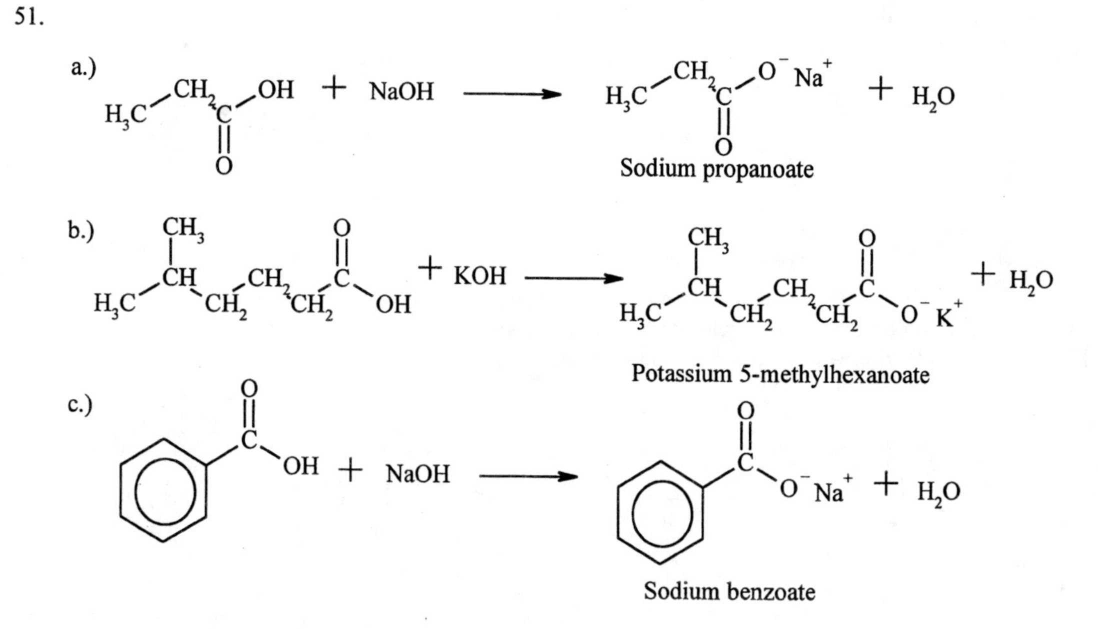

53. a. The carboxylic acid is methanoic acid (formic acid) and the alcohol is ethanol.
    b. The carboxylic acid is benzoic acid and the alcohol is methanol.

55.

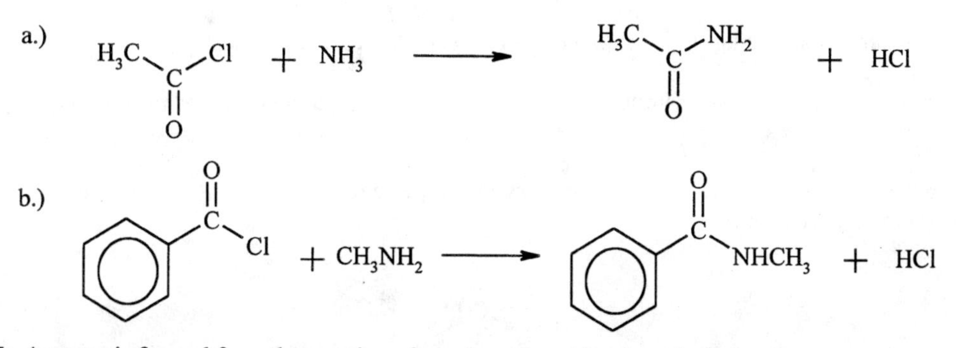

57. An ester is formed from the reaction of a carboxylic acid and an alcohol.

59. a. Cyclopentyl ethanoate
    b. Pentyl propanoate
    c. Propyl methanoate
    d. Methyl benzanoate

61. a. Isopropyl butanoate
    b. Methyl pyruvate

63.

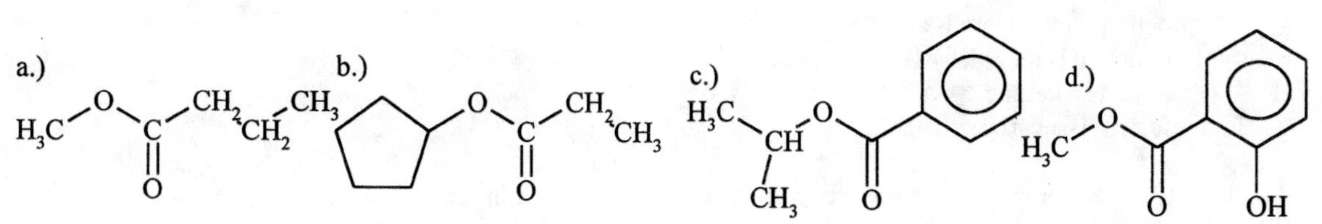

65. a. Methanamide or formamide
    b. Benzamide

67.

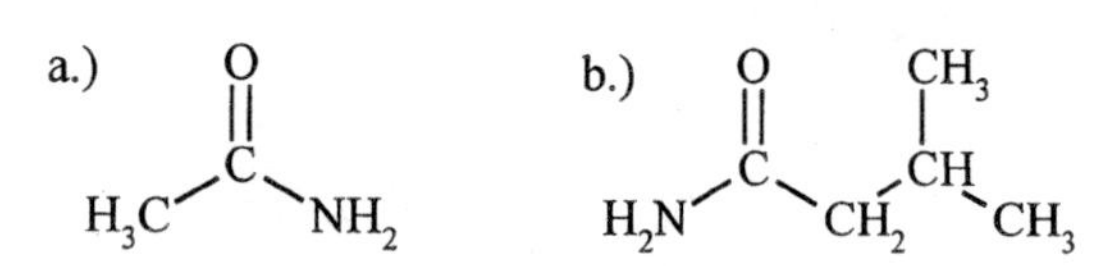

69. a. The hydrolysis of propanamide will yield propanoic acid and ammonia.
    b. The hydrolysis of benzamide will yield benzoic acid and ammonia.
    c. The hydrolysis of nylon will yield a diacid and a diamine.

71.

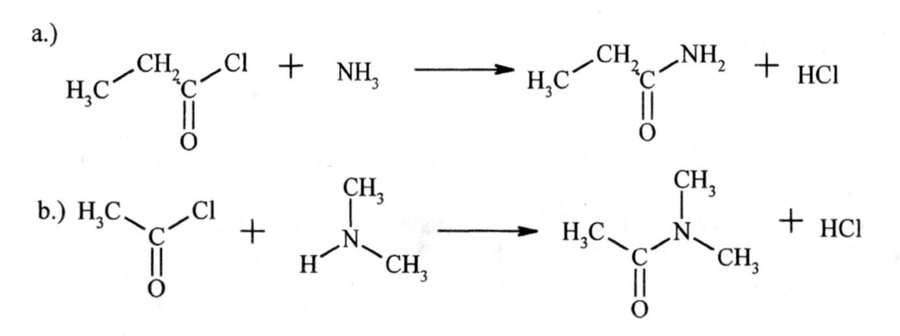

73. The functional group for an acid chloride is a chlorine atom attached to a carbonyl group.

75. The difference between a thiol ester and an ester is that the carbonyl carbon atom is attached to sulfur in the thiol ester and oxygen in an ester.

77. The reactant had to have been an aldehyde because there is a hydrogen atom attached to the carbon atom that was once the carbonyl carbon atom.

79.

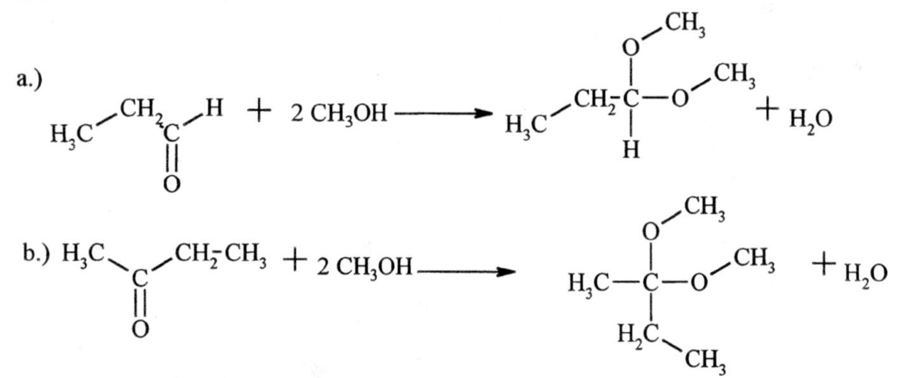

81. a. 1-bromoethyl methanoate is an ester.
    b. 4-chlorobutanal is an aldehyde.
    c. 3-pentanone is a ketone.
    d. Butanoic acid is a carboxylic acid.

83. Alkalosis is not as common as acidosis because acids are far more commonly encountered than bases. Your body is better able to adapt to the ingestion of acids than bases. One cause of alkalosis is the ingestion of poisons like illicit drugs. Many street drugs are prepared by amateurs that will cut the pure drug with chemicals

like Draino that are strong bases. Ingestion of a strong base is not able to be handled well by your body, and alkalosis can result.

Sample Test:

1. What is the difference between a carbon-oxygen double bond and a carbon-carbon double bond?
   a. The carbon-oxygen double bond is nonpolar.
   b. The carbon-carbon double bond is nonpolar.
   c. The carbon-oxygen double bond is harder to break than the carbon oxygen double bond.
   d. There is no difference between the two bonds.

2. Choose a name that fits the following aldehyde.

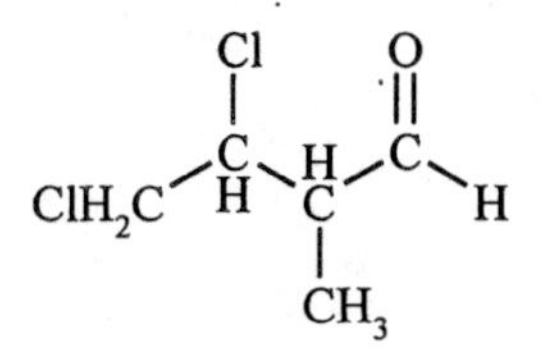

   a. 1,2-dichloro-3-methylbutanal
   b. 3,4-dichloro-2-methylpropanal
   c. 3,4-dichloro-2-methylbutyraldehyde
   d. 3,4-dichloromethylbutanal

3. Use the IUPAC rules to name glucose.

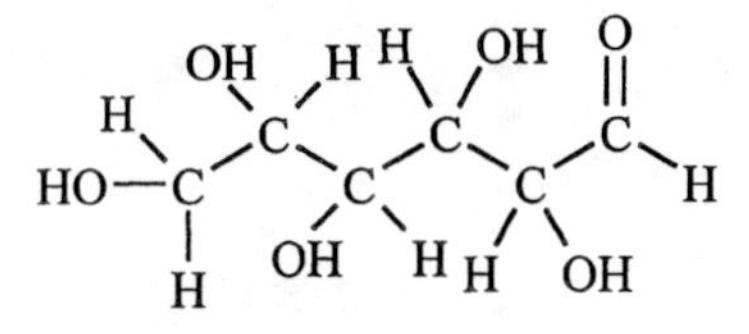

   a. Pentahydroxyhexanal
   b. Pentahydroxyhexeraldehyde
   c. 1,2,3,4,5-pentahydroxyhexanal
   d. 2,3,4,5,6-penatahydroxyhexanal

4. What name would apply to the following molecule?

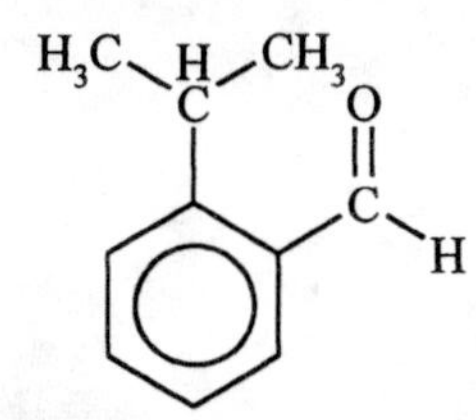

   a. Isopropylbenzaldehyde
   b. o-isopropylbenzaldehyde
   c. 1-isopropylbenzaldehyde
   d. More than one correct response.

5. The oxidation of butyraldehyde will yield what major organic product?
   a. 1-butanol
   b. 2-butanol
   c. Butanoic acid

d. A hemiacetal.

6. What will be the structure of the product if isopropanol is reacted with 2-methylbutanal?

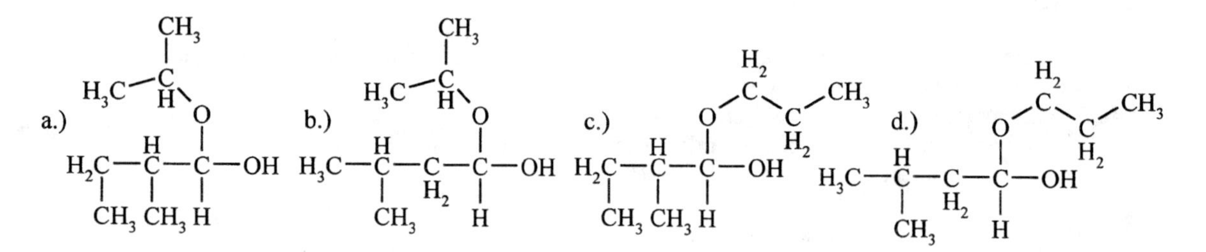

7. What is the name of the ketone drawn below?

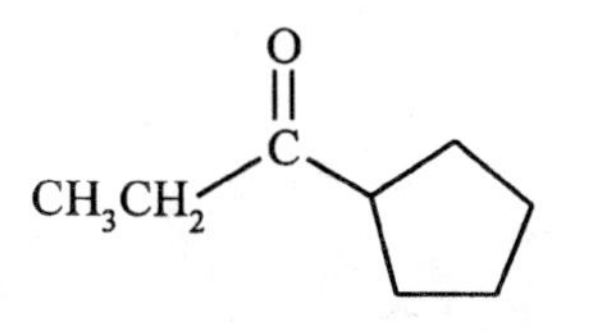

a. Ethylcyclopentanone
b. Ethyl cyclopentyl ketone
c. Cyclopentyl ethyl ketone
d. 3-cyclohexyl-3-propanone

8. What is the correct name for the following molecule?

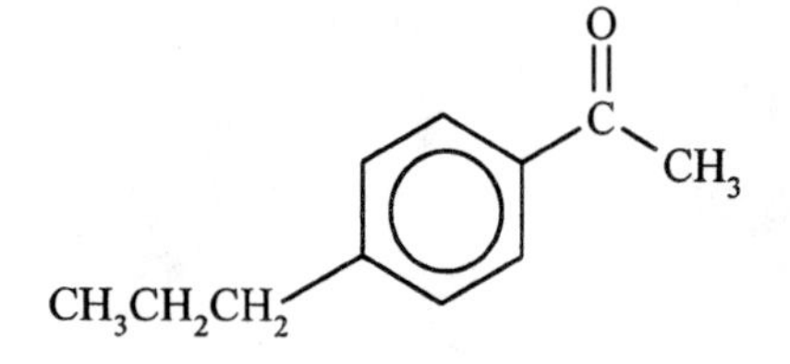

a. Methyl propylphenyl ketone
b. 4-isopropyl benzophenone
c. p-isopropyl acetophenone
d. p-propyl acetophenone

9. What were the reactants that produced the following molecule?

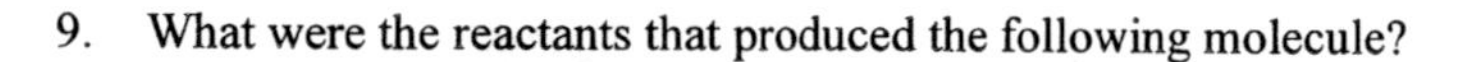

CH_3 (on O)
$H_3C—C—O—CH_2CH_2CH_3$
CH_3

a. One mole of acetone and two moles of methanol.
b. One mole of acetone and two moles 1-propanol
c. One mole of acetone, one mole of methanol, and one mole of ethanol.
d. One mole of acetone, one mole of methanol, and one mole of propanol.

10. What is the name of the following molecule?

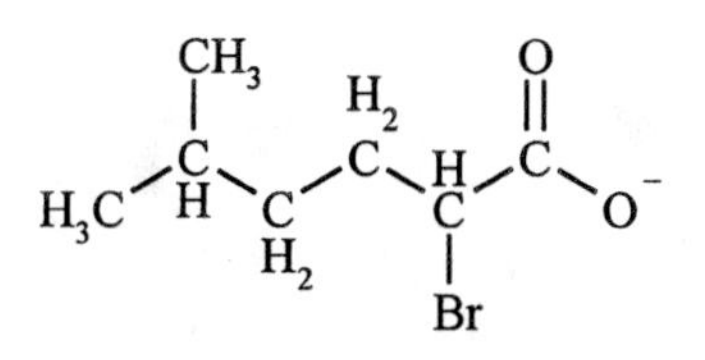

a. 5-methyl-2-bromohexanoic acid
b. δ-methyl-β-bromohexanoate
c. ε-methyl-α-bromohexanoate
d. δ-methyl-α-bromohexanoate

11. Which compound would you expect to be the most soluble in water?
    a. Acetone b. Methanol c. Heptanoic acid d. p-isopropylbenzoic acid

12. What is the name of the compound that forms when 2-methylbenzoic acid reacts with potassium hydroxide?
    a. Potassium o-methylbenzoate
    b. Potassium m-methylbenzoate
    c. Potassium 2-methylbenzoic acid
    d. 2-methylhydroxybenzoic acid

13. How does the following compound form?

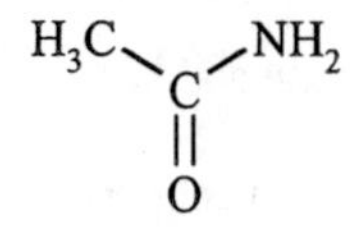

a. The reaction of acetic acid with ammonia.
b. The reaction of acetic acid with a primary amine.
c. The reaction of an acid chloride with ammonia.
d. The reaction of an acid chloride with a primary amine.

14. How does the following compound form?

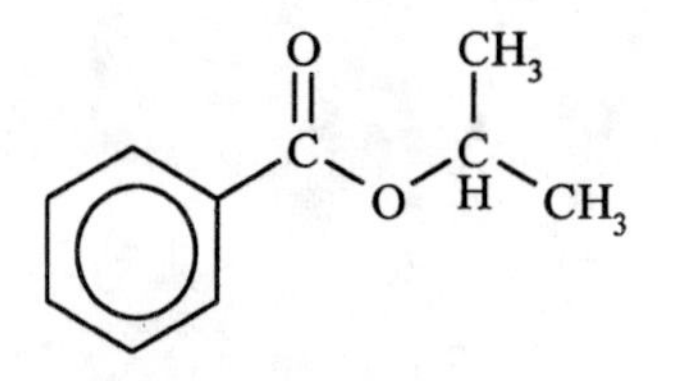

a. The reaction of benzoic acid with propanol.
b. The reaction of benzoic acid with isopropanol.
c. The reaction or benzaldehyde with isopropanol.
d. The reaction of benzophenone with isopropanol.

15. What class of compound will be formed from the reaction of an acid chloride with an amine?
    a. Carboxylic Acid b. Ester c. Amide d. Anhydride

16. What is the name of the ester formed from the reaction of ethanol with benzoic acid?

    a. Phenyl ethanoate b. Ethyl phenoate c. Ethyl benzoate d. Benzyl ethanoate

17. When aspirin is ingested, it is broken down by acids in your stomach. Hydrolysis of aspirin will give what product(s)?

Aspirin

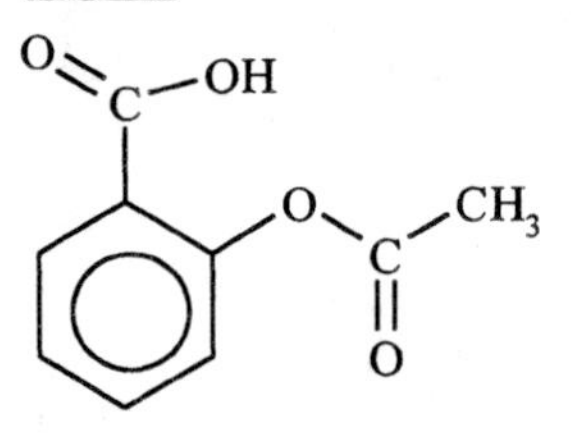

a. Benzoic acid
b. m-hydroxy benzoic acid
c. Acetic acid
d. More than one correct response.

18. What is the name of the product produced in the reaction of the acteyl chloride with diethyl amine?
    a. N,N-diethyl ethanamide
    b. N-ethyl butanamide
    c. N,N-triethanamide
    d. N-acetyl diethyl amide

19. What is(are) the product(s) of the hydrolysis of 2-methyl propanamide?
    a. Ammonia   b. Isopropanol   c. N-methyl amine   d. 2-methyl propanol

20. What type of compound will be produced when two carboxylic acids react with one another?

    a. Aldehyde   b. Ketone   c. Acid chloride   d. Acid anhydride

Answers to Sample Test:

1. b
2. c
3. d
4. b
5. a
6. a
7. c
8. d
9. d
10. d
11. b
12. a
13. c
14. b
15. c
16. c
17. c
18. a
19. a
20. d

Chapter 12: Carbohydrates

Chapter Objectives:

1. To identify and distinguish among the different classes of carbohydrates.
2. To recognize the common monosaccharides.
3. To understand the different forms of carbohydrates and how they interact.
4. To become familiar with carbohydrates containing two to ten monomers.
5. To recognize and become familiar with common polysaccharides.

Key Terms: The key terms are listed in the order in which they are encountered in the chapter.

Biochemistry. The study of the chemistry of living organisms.

Biomolecules. The compounds of living organisms that include carbohydrates, lipids, proteins, and nucleic acids.

Macromolecule. A very large molecule.

Carbohydrates. Polyhydroxy aldehydes or ketones, or derivatives of these compounds.

Monosaccharides. Simple sugars that have the general formula $C_nH_{2n}O_n$ that often serve as the building blocks for larger carbohydrates.

Oligosaccharides. Carbohydrates that contain between two and ten monosaccharides covalently linked together.

Polysaccharides. Carbohydrates that contain more than ten monosaccharide subunits linked together by covalent bonds.

Aldose. A monosaccharide that contains an aldehyde group (RCHO).

Ketose. A monosaccharide that contains a ketone (RCOR').

Levo- (L). The prefix L- is used to designate that the position of the hydroxyl group attached to the highest priority chiral carbon atom in a carbohydrate is to the left when the carbonyl group is drawn up.

Dextro- (D). The prefix D- is used to designate that the position of the hydroxyl group attached to the highest priority chiral carbon atom in a carbohydrate is to the right when the carbonyl group is drawn up.

Highest priority chiral carbon. In a simple sugar, the carbonyl group is drawn up and the highest priority chiral carbon is the chiral center that is furthest away from the carbonyl carbon.

Intramolecular hemiacetal. The hemiacetal formed when the hydroxyl group of the highest priority chiral center bonds to the carbonyl carbon atom of the same pentose or hexose.

Furanose. A cyclic monosaccharide that contains five atoms in a heterocyclic ring system as the result of intramolecular hemiacetal formation of an aldopentose or a ketohexose.

Pyranose. A cyclic monosaccharide that contains six atoms in a heterocyclic ring system as the result of intramolecular hemiacetal formation of an aldohexose.

Anomeric carbon. The carbon atom of a monosaccharide that was the carbonyl carbon atom prior to intramolecular hemiacetal formation.

Anomers. The D- and L- isomers that are the result of intramolecular hemiacetal formation.

Hyperglycemia. The condition that exists when blood sugar levels exceed their normal range of 60 mg/dl to 100 mg/dl.

Hypoglycemia. The condition that exists when blood sugar levels are below their normal range of 60 mg/dl to 100 mg/dl.

Optically active. A compound that rotates a beam of plane polarized light due to the presence of a chiral center.

Reducing agent. A substance that oxidized thus causing the reduction of another substance.

Reducing sugar. A sugar that contains a carbonyl group that can be oxidized. All monosaccharides are reducing sugars.

Glycoside. The carbohydrate equivalent of an acetal that is created when a cyclic carbohydrate (hemiacetal form) reacts with an alcohol.

Glycosidic bond. The bond that forms between the anomeric carbon atom of a cyclic sugar and the oxygen atom of an alcohol.

Glucoside. The glycosidic bond that forms between the anomeric carbon atom of the cyclic form of glucose and the oxygen atom of an alcohol.

Fructoside. The glycosidic bond that forms between the anomeric carbon atom of the cyclic form of fructose and the oxygen atom of an alcohol.

Galactoside. The glycosidic bond that forms between the anomeric carbon atom of the cyclic form of galactose and the oxygen atom of an alcohol.

Residue. The monomer units that make up a polymer.

Disaccharide. An oligosaccharide that contains two monosaccharides linked by a glycosidic bond.

Lactose. The reducing disaccharide found in milk that consists of galactose and glucose linked by a $\beta(1\rightarrow4)$ glycosidic bond.

Sucrose. The non-reducing disaccharide found in plants and flowers that consist of fructose and glucose linked by an $\alpha,\beta(1\rightarrow2)$ glycosidic bond.

Maltose. The reducing disaccharide found in germinating grain that consists of two molecules of glucose linked by an $\alpha(1\rightarrow4)$ glycosidic bond.

Erythrocytes. Human red blood cells.

Polysaccharides. A carbohydrate polymer made up of monosaccharide monomer units linked together by glycosidic bonds.

Starch. The major storage polysaccharide in plants that is made of D-glucose molecules linked by $\alpha(1\rightarrow4)$ and $\alpha(1\rightarrow6)$ glycosidic bonds.

Complex carbohydrates. Polysaccharides that are used by plants or animals to store energy.

Amylose. One type of polymer found in starches that has glucose residues connected in a linear chain by $\alpha(1\rightarrow4)$ glycosidic bonds.

Amylopectin. The larger branched polymer found in starches that has glucose residues connected by both $\alpha(1\rightarrow4)$ and $\alpha(1\rightarrow6)$ glycosidic bonds.

Glycogen. A polymer of $\alpha(1\rightarrow4)$ linked subunits of glucose with $\alpha(1\rightarrow6)$ branches that serves as the major storage polysaccharide in animals.

Glycogen phosphorylase. An enzyme that catalyzes the release of individual phosphorylated glucose molecules from the ends of the chains of glycogen.

Cellulose. A linear D-glucose polymer linked by $\beta(1\rightarrow4)$ glycosidic bonds that serves as the structural material for plants.

Chitin. The polysaccharide found in the exoskeleton of arthropods that has the same structure as cellulose with the hydroxyl group attached to carbon number two replaced by an amide group.

Lignin. A structural polymer that is supported by cellulose to provide structure and protections for plants.

Fiber. Indigestible plant matter that consists primarily of cellulose.

Hyaluronic acid. A polysaccharide found in the connective tissue of higher animals that serves to absorb shock and lubricate bone surfaces.

Heparin. A sulfate rich polysaccharide that is a powerful anticoagulant.

Chapter 12 Objective Details:

a. Classes and names of carbohydrates. Carbohydrates are polyhydroxy aldehydes or ketones or derivatives of these compounds that have the general formula $C_nH_{2n}O_n$. A single carbohydrate is called a monosaccharide. Monosaccharides serve as building blocks for oligosaccharides (2-10 monosaccharides) and polysaccharides (10 or more monosaccharides).

b. Monosaccharides. Monosaccharides are simple sugars that are classified by the number of carbon atoms in their parent chain, or by the functional group they contain. The number of carbon atoms in a monosaccharide is specified by the prefixes tri-, tetr-, pent-, and hex- to indicate 3, 4, 5, and 6 carbon atoms, respectively. The suffix –ose is used to indicate that the molecule is a carbohydrate. The two main classes of monosaccharides are aldoses that contain an aldehyde functional group and ketoses that contain a ketone functional group. The functional group of the carbohydrate and the number of carbon atoms can be designated by combining the two systems of classification. For example an aldohexose is a sugar (suffix – ose) that contains six carbon atoms (hex-) and an aldehyde functional group (aldo-).

The prefixes D- and L- are used prior to the name of a carbohydrate to indicate whether the highest priority chiral carbon atom has the hydroxyl functional group on the right or the left side, respectively. The assumption is that the structure is drawn with the carbonyl group toward the top of the paper or in the up position. Double bonds are flat and have restricted or hindered rotation. In the intramolecular hemiacetal reaction, the hydroxyl group attached to the highest priority chiral carbon can bond to the bottom side of the carbonyl carbon or the topside of the carbonyl carbon atom. The result is the formation of either the beta or alpha form of the sugar. It should be noted that the only carbohydrates that are commonly encountered in either the furanose or pyranose forms are the pentoses and hexoses.

Example: How do you determine D-, L-, α-, or β-?

Solution:

1.) Start with the straight chain form of the carbohydrate you are interested in drawing.
2.) Draw the structure so the carbonyl group is toward the top of the paper and look for the chiral center that is furthest from the carbonyl group. Is the hydroxyl group to the right (D-) or left (L-) of the main chain?

3.) Intramolecular hemiacetal formation will result in a heterocyclic five-membered (furanose) or six-membered (pyranose) ring. Always draw the cyclic forms of the carbohydrates with the oxygen atom of the pyranose ring in the top right corner and the oxygen atom of the furanose ring in the top center of the structure. If the linear form of the carbohydrate was D-, the $-CH_2OH$ group goes up and if the linear form of the carbohydrate was L-, the $-CH_2OH$ group goes down.
4.) The key is to remember that you are reacting an alcohol with either an aldehyde or a ketone. That means that the oxygen atom in the ring is from the hydroxyl group of the highest priority chiral carbon. Make sure you label your atoms starting from the top with the number 1 and going to the last carbon atom. The carbonyl oxygen atom is going to be converted to a hydroxyl group that will either go up or down. Both isomers can form. *If the hydroxyl group forms to the same side of the ring as the $-CH_2OH$, the β-isomer has been formed. If the hydroxyl group forms to the opposite side of the ring as the $-CH_2OH$, the α-isomer has been formed.*
5.) Draw the structure until it makes sense!

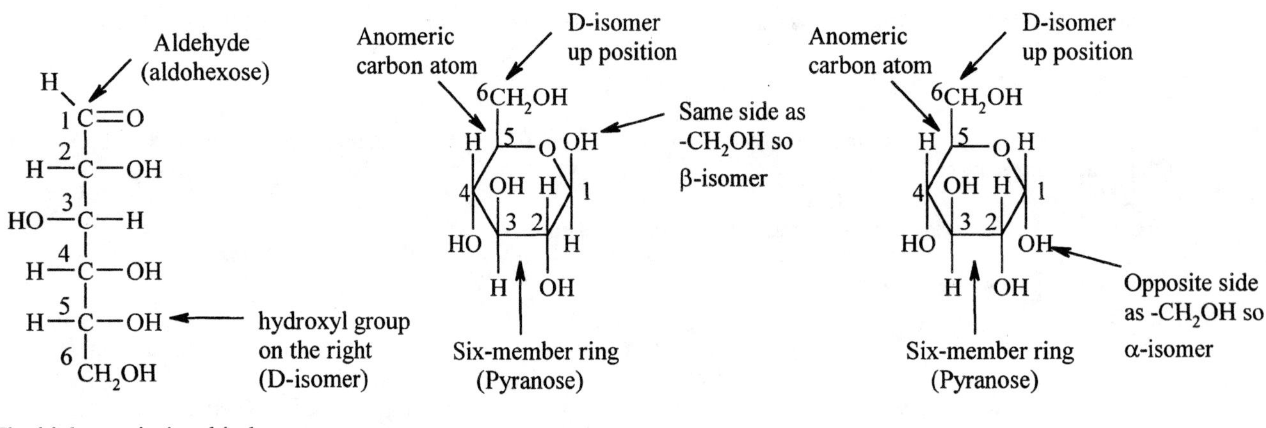

The highest priority chiral center (C-5) has the hydroxyl group on the right side. This is the D-isomer of glucose.

The common monosaccharides to become familiar with are the trioses D-glyceraldehyde and dihydroxyacetone, pentoses ribose and 2-deoxyribose, and hexoses glucose, fructose and galactose. Glyceraldehyde is the smallest aldose and dihydroxyacetone is the smallest ketose. Ribose and 2-deoxyribose are key components to ribonucleic acid (ribo = ribose) or RNA and deoxyribonucleic acid (deoxyribo = 2-deoxyribose) or DNA. Glucose is blood sugar and galactose is commonly converted to glucose during normal metabolism. Fructose is a ketohexose that is a primary component of sucrose or table sugar and honey.

c. Properties and reactions of sugars. Sugars are polyhydroxy aldehydes and polyhydroxy ketones. They are water soluble in the monomeric and oligomeric forms. As the polymer chain increases in the case of polysaccharides, the sugars are no long water-soluble. Sugars can rotate plane-polarized light in a clockwise direction (+) or a counterclockwise direction (-). The ability of a compound to rotate plane-polarized light is dependent upon the chirality of the molecule.

Monosaccharides can be oxidized, thus causing the reduction of another agent. It is therefore the case that all monosaccharides are reducing sugars. In addition to oxidation-reduction reactions, cyclic carbohydrates are in the hemiacetal form and can react with alcohols to form acetals. The bond that forms between the hemiacetal and the alcohol is called a glycosidic bond.

d. Oligosaccharides. Oligosaccharides are carbohydrates containing two to ten monosaccharides bonded together by glycosidic linkages. The most common oligosaccharides are the disaccharides that consist of two residues or monomer units. The disaccharides most commonly encountered are lactose (milk sugar), sucrose (table sugar), and maltose (grain sugar). Lactose is a reducing sugar that consists of galactose and glucose linked by a β(1→4) glycosidic bond. The beta one to four glycosidic bond is between the hydroxyl group, attached to carbon number one in galactose, that is in the beta position bonded to the four position of the pyranose form of glucose. Sucrose is a not a reducing sugar, and consists of fructose and glucose linked by an α,β(1→2) glycosidic bond. Maltose is a reducing sugar that consists of two molecules of glucose linked by a α(1→4) glycosidic bond.

e. Polysaccharides. Polysaccharides are carbohydrates consisting of ten or more monomer units connected by glycosidic bonds. Animals store carbohydrates as glycogen, which is a D-glucose polymer joined by α(1→4), and α(1→6) glycosidic linkages. Starch is the carbohydrate storage form for plants. Starch can exist as amylose, a linear polymer of repeating D-glucose units linked by α(1→4) glycosidic bonds. Alternatively, starch can exist as amylopectin. Amylopectin is a D-glucose polymer joined by α(1→4), and α(1→6) glycosidic linkages that is less branched and larger than glycogen. Cellulose is the D-glucose polymer found in the cell walls of plants that is held together by β(1→4) glycosidic bonds. The beta bonds are more compact and less exposed to hydrolysis than alpha bonds. For this reason, humans are unable to breakdown cellulose. Chitin is the same as cellulose with the exception that the hydroxyl group of the second carbon atom of glucose is replaced by an amide. Chitin as a polymer is very solid and serves as the exoskeleton to arthropods. Hyaluronic acid and heparin are polysaccharides that have more complex repeating units that serve as protectors of joints and bones and as anticoagulants, respectively.

Solutions to Odd Numbered Problems:

1. a. A monosaccharide is a simple sugar that has the general formula $C_nH_{2n}O_n$ that often serve as a building block for larger carbohydrates.
   b. An oligosaccharide is a carbohydrate that contains between two and ten monosaccharides covalently linked together.
   c. A polysaccharide is a carbohydrate that contains more than ten monosaccharide subunits linked together by covalent bonds.

3. The origin of the word carbohydrate comes from the observation that dehydration of a carbohydrate yielded only carbon. It was assumed that these compounds were the hydrates of carbon and thus the term "carbo" for carbon and "hydrate" for containing water came to be accepted.

5. a. A aldotriose is a carbohydrate containing an aldehyde functional group and three carbon atoms.
   b. A ketoheptose is a carbohydrate containing a ketone functional group and seven carbon atoms.
   c. An aldopentose is a carbohydrate containing an aldehyde functional group and five carbon atoms.

7. The smallest aldose is glyceraldehyde and the smallest ketose is dihydroxyacetone.

9. a. L-aldotetrose
   b. D-ketopentose

11.

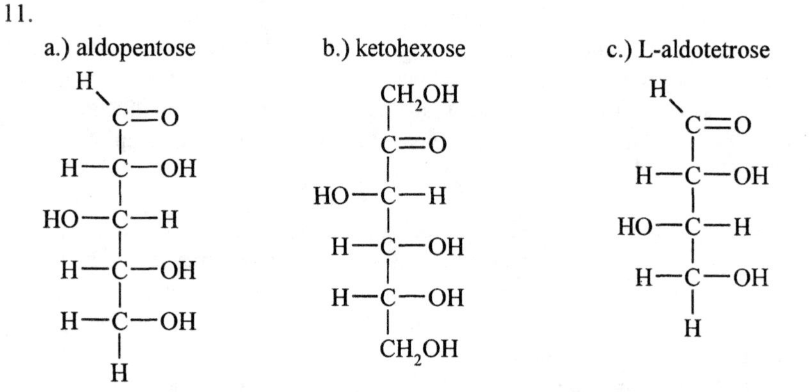

13. To convert a D-aldohexose to the L-aldohexose, switch the hydroxyl group attached to carbon number 5 from the right hand side to the left hand side of the structure.

15. a. D-ribose
    b. L-glyceraldehyde

17. a. Oligosaccharide
    b. Monosaccharide
    c. Oligosaccharide
    d. Polysaccharide

19. Ribose and 2-deoxyribose are the same except the hydroxyl group on the second carbon of ribose is replaced by a hydrogen atom in 2-deoxyribose.

21. The anomeric carbon atom is the carbon atom that originated from the carbonyl group. The anomeric carbon atom for an aldopentose would be carbon number one. The anomeric carbon atom for a ketohexose would be carbon number two.

23. a. This is the representation of a cyclic hemiacetal where the anomeric carbon atom is in the two position. The two position is immediately left of the oxygen atom in the ring.
    b. This structure is not a cyclic hemiacetal.

25. The furanose form of a monosaccharide contains a five-member heterocyclic ring where the hetero-atom is oxygen.
    The pyranose form of a monosaccharide contains a six-member heterocyclic ring where the hetero-atom is oxygen.

27. Monosaccharides and disaccharides are able to form sufficient hydrogen-bonding interactions with water to make them soluble.

29. a. D-fructose contains three chiral centers.
    b. A ketohexose would have eight stereoisomers.

31. A

33. Both galactose and lactose contain a hemiacetal that can be oxidized. The oxidation of these sugars makes them reducing agents and thus reducing sugars.

35. All monosaccharides are reducing sugars.

37. a. This is an alpha fructoside.
    b. This is a alpha glucoside.
    c. This is a beta-2-deoxyriboside.

39. C

41. a. Hydrolysis of sucrose will yield D-glucose and D-fructose.
    b. Hydrolysis of maltose will yield two molecules of D-glucose.
    c. Hydrolysis of lactose will yield D-galactose and D-glucose.

43. One of the two structures of maltose is drawn indicating the acetal and the glycosidic bond.

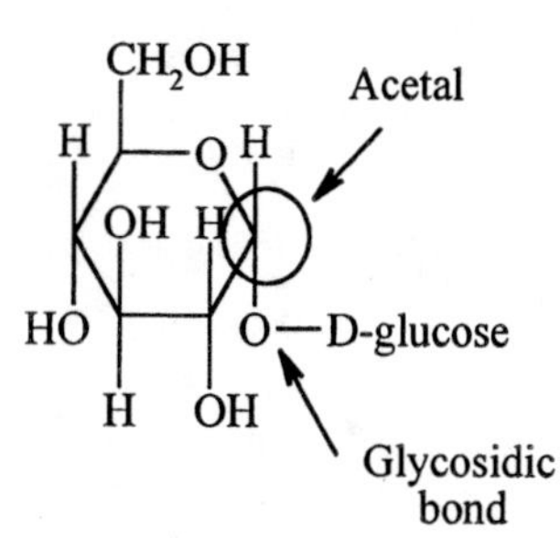

45. The polysaccharides that yield only D-glucose when hydrolyzed are amylose, amylopectin, glycogen, and cellulose.

47. Amylose is linear and amylopectin is branched.

49. Glycogen is used for carbohydrate storage. It is highly branched, which allows for several molecules of glucose to by hydrolyzed at the same time. The availability of glucose helps to maintain normal glucose levels in the body.

51. Humans lack the enzymes necessary to digest cellulose. Cattle contain a microorganism that can make the enzymes necessary to hydrolyze the beta linkages in cellulose.

53. Heparin is a polymer containing a two-sugar repeating unit that is abundant with sulfate groups. Heparin acts as an anticoagulant preventing the clotting of blood.

55. a. Carbohydrates are polyhydroxy aldehydes or polyhydroxy ketones.
    b. A pentose is a carbohydrate containing five carbon atoms.
    c. An aldose is a carbohydrate containing the aldehyde functional group.
    d. A ketose is a carbohydrate containing the ketone functional group.
    e. A ketohexose is a carbohydrate containing the ketone functional group and six carbon atoms.
    f. The L-family are chiral sugars where the highest priority chiral center has the hydroxyl group on the left assuming the structure is drawn with the carbonyl towards the top of the page.
    g. A furanose ring is a heterocyclic ring system containing five atoms of which one is oxygen.
    h. The anomeric carbon atom of a cyclic carbohydrate is the carbonyl carbon atom of the straight chain form.
    i. Anomers are the two isomers that are the result of intramolecular hemiacetal formation in a pentose or hexose.
    j. A glycosidic bond is the bond formed between the hemiacetal of a carbohydrate and an alcohol.
    k. Monosaccharides are carbohydrates with the general formula $C_nH_{2n}O_n$.
    l. A disaccharide contains two monosaccharides linked by a glycosidic bond.
    m. A polysaccharide is many monosaccharides linked by glycosidic bonds.
    n. Glycogen is a polymer that serves as the major carbohydrate storage unit for animals.
    o. Cellulose is a polymer that serves as the major structural component in plant cell walls.
    p. Amylose is a form of starch that exists as a linear polymeric chain of D-glucose units.
    q. A reducing sugar is a sugar that can be oxidized.
    r. Lactose is milk sugar, a disaccharide made of glucose and galactose.

57. a. Reducing
    b. Reducing
    c. Not a reducing sugar
    d. Reducing (Because of the highly branched nature of glycogen, it may give a positive test.)
    e. Not a reducing sugar
    f. Reducing

59. Assuming that complex carbohydrates are fully digestible, use the example on page 403 of Chapter 13 to solve this problem.

    a. $11 \text{ g sugar} \times \frac{4 \text{ cal}}{1 \text{ g}} = 44 \text{ cal}$

    $15 \text{ g complex carbo} \times \frac{4 \text{ cal}}{1 \text{ g}} = 60 \text{ cal}$

    total = 44 cal + 60 cal = 104 cal

    b. $\text{Percent} = \frac{60 \text{ cal complex carbo}}{104 \text{ cal total}} \times 100 = 58\%$

Sample Test:

1. Which of the following molecules would be a carbohydrate?

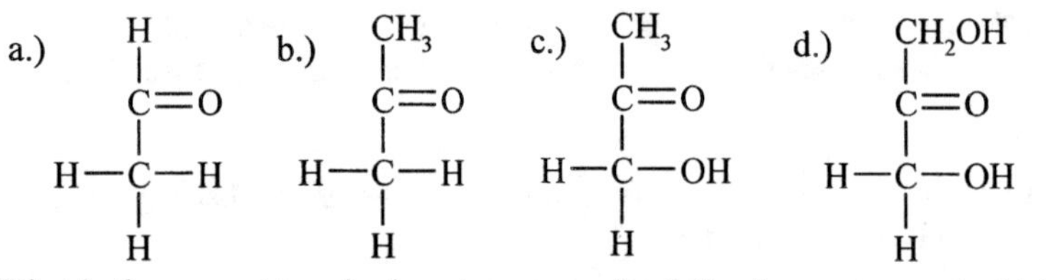

2. What is the correct terminology to express the following monosaccharide?

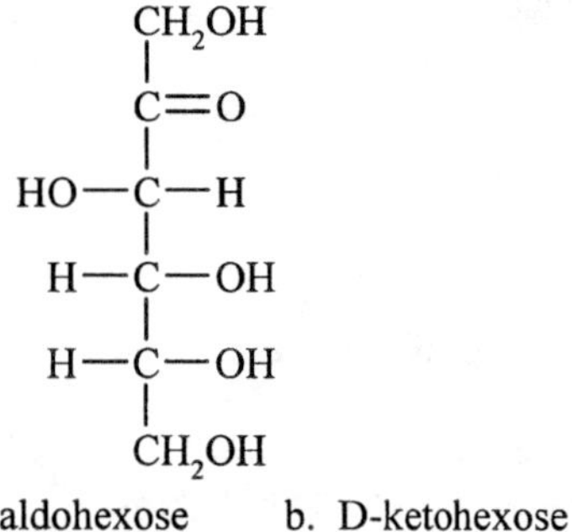

a. D-aldohexose b. D-ketohexose c. L-aldohexose d. L-ketohexose

3. What is the difference between an aldopentose and a ketohexose?
   a. The aldopentose will cyclize to the furanose form and the ketohexose will cyclize to the pyranose form.
   b. The anomeric carbon atom of an aldopentose is number two and for a ketohexose it is number one.
   c. The aldopentose contains five carbon atoms and the ketohexose contains six carbon atoms.
   d. The functional group for the aldopentose is an alcohol and a ketone for the ketohexose.

4. Identify the sugar by α- or β- isomer and D- or L- family.

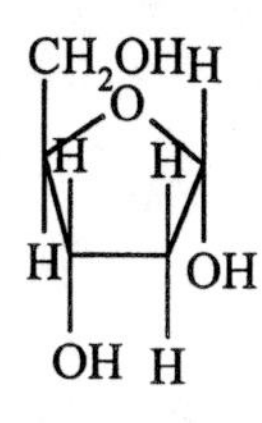

a. α-D-furanose   b. α-L-furanose   c. β-D-furanose   d. β-L-furanose

5. What is the difference between the D- and L- forms of glyceraldehyde?
   a. The position of substituents about carbon number one.
   b. The position of substituents about carbon number two.
   c. The position of substituents about carbon number three.
   d. The position of substituents about carbon number four.

6. Which pair of monosaccharides contains different numbers of carbon atoms?
   a. Fructose and ribose
   b. Ribose and deoxyribose
   c. Glucose and fructose
   d. Glyceraldehyde and dihydroxyacetone

7. Which of the following statements concerning chiral centers in specific monosaccharides is correct?
   a. Glyceraldehyde has two chiral center.
   b. Dihydroxyacetone has two chiral centers.
   c. Fructose has three chiral centers.
   d. Glucose has five chiral centers.

8. Which of the following monosaccharides will yield a furanose ring upon cyclization?
   a. Aldopentoses and ketopentoses.
   b. Aldopentoses and ketohexoses.
   c. Aldohexoses and ketopentoses.
   d. Aldohexoses and ketohexoses.

9. Which pair of carbohydrates are disaccharides?
   a. Glucose and fructose
   b. Lactose and amylose
   c. Sucrose and maltose
   d. Amylose and amylopectin

10. What carbohydrate will produce both galactose and glucose upon hydrolysis?
    a. Sucrose   b. Maltose   c. Lactose   d. Amylose

11. Which of the following is not a reducing sugar?
    a. Lactose   b. Sucrose   c. glucose   d. fructose

12. What is the major structural difference between cellulose and starch?
    a. The identity of the monosaccharide units present.
    b. The ring size of the monosaccharide units present.
    c. The linkages between the monosaccharide units present.
    d. The handedness (D- or L-) of the monosaccharide units present.

13. Of the following polysaccharides, which pair represents branched polysaccharides?
    a. amylopectin and glycogen
    b. cellulose and glycogen

c. cellulose and amylose
d. amylose and amylopectin

14. Which polysaccharide contains all β(1→4) glycosidic linkages?
a. Starch b. Cellulose c. Glycogen d. Amylopectin

15. Which of the following statements concerning carbohydrates is correct?
a. They are the most abundant type of biochemical substance in the human body.
b. They are all polyhydroxy aldehydes, polyhydroxy ketones, or substances that yield such upon hydrolysis.
c. They can exist in left-handed and right-handed forms, with the left-handed form being dominant in nature.
d. They all have a sweet taste and therefore are also called sugars.

16. Which of the following monosaccharides is an aldohexose?
a. Glyceraldehyde b. 2-Deoxyribose c. Fructose d. Galactose

17. Which of the following monosaccharides has more than three chiral centers?
a. Ribose b. Fructose c. Glucose d. Glyceraldehyde

18. Which of the following monosaccharides will form a pyranose ring upon cyclization?
a. Fructose b. Galactose c. Ribose d. Glyceraldehyde

19. Which of the following disaccharides contains a β(1→4) glycosidic linkage?
a. Lactose b. Sucrose c. Maltose d. No correct response.

20. Which of the following carbohydrates produces only glucose upon hydrolysis?
a. Maltose b. Starch c. Cellulose d. More than one correct response.

Answers to Sample Test:

1. d
2. b
3. c
4. a
5. b
6. a
7. c
8. b
9. c
10. c
11. b
12. c
13. a
14. b
15. b
16. d
17. c
18. b
19. a
20. d

Chapter 13: Lipids

Chapter Objectives:

1. To name and recognize the common fatty acids.
2. To draw and understand the biological significance of triacylglycerols.
3. To describe the structure and properties of polar lipids.
4. To recognize the structure and function of simple lipids.
5. To explain membranes and how they work.

Key Terms: The key terms are listed in the order in which they are encountered in the chapter.

Lipids. Biological molecules that are insoluble in water but soluble in organic solvents.

Fatty acids. Long-chain carboxylic acids.

Complex lipids. Lipids that contain long-chain carboxylic acids.

Simple lipids. Lipids that do not contain long-chain carboxylic acids.

Saturated fatty acids. Long-chain carboxylic acids that contain no carbon-carbon double or triple bonds.

Unsaturated fatty acids. Long-chain carboxylic acids that contain one or more carbon-carbon double bonds.

Monounsaturated fatty acid. Long-chain carboxylic acids that contain only one carbon-carbon double bond.

Polyunsaturated fatty acid. Long-chain carboxylic acids that contain two or more double bonds.

Amphipathic molecule. A molecule that contains both polar and nonpolar ends.

Triacylglycerols. Nonpolar complex lipids consisting of the fatty acid triester of glycerol that serve as energy reserves.

Monoacylglycerol. A molecule of glycerol containing one fatty acid chain bonded by an ester linkage.

Diacylglycerol. A molecule of glycerol containing two fatty acid chains bonded by ester linkages.

Triglycerides. The same as a triacylglycerol; a compound containing three fatty acids esterified to glycerol.

Fats. Triacylglycerols that are solid at room temperature.

Oils. Triacylglycerols that are liquids at room temperature.

Iodine number. The grams of iodine required to react with all of the carbon-carbon double bonds in 100 g of sample.

Rancid. The bad odor and taste associated with the products of oxidation of unsaturated triacylglycerols.

Essential fatty acids. Fatty acids that are not synthesized in the in the body, but they are needed by the body to produce other necessary biomolecules.

Cell membrane. A sheet-like bilayer structure that surrounds a cell and divides the interior of the cell.

Phospholipids. Lipids that contain phosphorus.

Phosphatidic acid. A molecule consisting of glycerol with two fatty acids connected by ester bonds and a phosphoric acid connected to the third hydroxyl group of glycerol also by an ester linkage.

Phosphoacylglycerols. A molecule of phosphatidic acid with an alcohol connected to the phosphate group through an ester linkage.

Phosphatidyl choline. A phosphoacylglycerol where the alcohol bonded to the phosphate group is choline.

Phosphatidyl ethanolamine. A phosphoacylglycerol where the alcohol bonded to the phosphate group is ethanolamine.

Polar lipids. Amphipathic molecules that contain a polar head region and a nonpolar tail region.

Sphingolipids. Amphipathic lipids that contain the base sphingosine, a fatty acid, and one or more other molecules.

Sphingomyelin. A sphingolipid that contains the base sphingosine, a fatty acid, and the alcohol choline bonded to a phosphate group bonded to sphingosine.

Cerebroside. A sphingolipid that contains the base sphingosine, a fatty acid, and a carbohydrate.

Glycolipid. Lipids that contain a carbohydrate as part of their structure.

Steroids. Compounds that contain four rings fused in a pattern termed the steroid nucleus.

Steroid nucleus. A network of three six-member rings joined to a five-member ring in a pattern consistent and characteristic of all steroids.

Cholesterol. The most abundant amphipathic steroid found in animals that keeps membranes fluid and serves as a starting place for the synthesis of many other steroids.

Atherosclerosis. The condition that exists on account of cholesterol deposits forming on the interior surface of arteries.

Steroid hormones. Molecules that are produced in one part of the body and are then circulated in blood to the rest of the body, where they bind to specific target cells and exert influence over the activity of these cells.

Sex hormones. Steroid hormones that determine the secondary sexual characteristics of males and females.

Androgens. Male sex hormones that include testosterone.

Progesterone. A female sex hormone that is very similar in structure to testosterone.

Estrogens. Common female sex hormones that contain an aromatic ring as part of the steroid nucleus.

Adrenocorticoid hormones. Steroid hormones that are produced by the cortex of the adrenal gland to regulate aspects of metabolism.

Hydrocortisone. An adrenocorticoid hormone also known as cortisol that is a powerful anti-inflammatory agent.

Bile salts. Amphipathic steroids that are synthesized from cholesterol for the purpose of triacylglycerol digestion.

Anabolic steroids. Hormones that stimulate the body's reactions related to synthesis and growth.

Antipyretic. A compound that reduces fever.

Prostaglandins. A type of lipid synthesized from polyunsaturated fatty acids that is thought to be responsible for the body's regulatory properties of fever and pain.

Leukotrienes. A type of lipid synthesized from polyunsaturated fatty acids that is thought to be responsible for allergic reactions and inflammation responses.

Micelle. A spherical structure that has a nonpolar, hydrophobic environment in the core and a polar, aqueous environment on the surface.

Hydrophobic interactions. Intramolecular interactions between nonpolar molecules and bring them together in a stable arrangement in an aqueous medium.

Hydrophilic interactions. The intramolecular interaction between polar molecules that their aqueous environment that stabilize the system.

Bilayer. A structure that forms when amphipathic molecules are put into water.

Liposomes. Bilayers that close back on themselves to form a continuous bilayer surrounding a core of water.

Fluid mosaic model. The current model for cellular membranes that has a lipid bilayer with proteins and cholesterol in the interior hydrophobic portion, and carbohydrates and hydrophilic protein portions on the exterior.

Simple diffusion. The movement of molecules across a membrane from regions of higher concentration to regions of lower concentration.

Facilitated diffusion. The transport of ions or polar molecules through a membrane with the assistance of proteins containing polar channels between the inside and outside of the membrane.

Active transport. The transport of ions and polar molecules through a protein, across a membrane, against a concentration gradient, and with the expenditure of energy.

Vitamin A. A fat-soluble vitamin that serves a role in the detection of light in the retina of the eye.

Vitamin D. A fat-soluble vitamin that serves a role in calcium uptake and use and in proper bone mineralization.

Vitamin E. A fat-soluble vitamin that serves a role in reducing the oxidation of double bonds thus making it an antioxidant.

Vitamin K. A fat-soluble vitamin that serves a role in the proper clotting of blood.

Tocopherols. Antioxidants including vitamin E that reduce the rate of oxidation of double bonds.

Lipoproteins. Proteins that bind lipids and transport them through the blood.

Chylomicrons. Lipoproteins that transport dietary fats and cholesterol from the intestine to the tissues of the body.

Low-density lipoproteins (LDLs). Lipoproteins that transport cholesterol and other lipids from the liver to the cells of the body where they are absorbed and used.

High-density lipoproteins (HDLs). Lipoproteins that transport cholesterol from body cells back to the liver.

Chapter 13 Objective Details:

a. Fatty acids. Fatty acids are long-chain carboxylic acids that can be either saturated (no double bonds), monounsaturated (1 double bond), or polyunsaturated (2 or more double bonds). Complex lipids contain fatty acids and simple lipids do not. Lipids are biological molecules that are soluble in organic solvents, but insoluble in water.

Fatty acids are amphipathic molecules that contain both a polar carboxyl group and a nonpolar hydrocarbon group. The London dispersion forces of fatty acids account for the observation that the longer the hydrocarbon chain of the fatty acid, the higher the melting point and the higher the boiling point of the compound. Unsaturated fatty acids have primarily cis- double bonds that interfere with the packing of the hydrocarbon chains. This inability to pack as tightly lowers both the melting point and boiling point for the unsaturated fatty acids. The higher the degree of unsaturation, the lower the melting point and boiling point. If you spend any time in the kitchen, you have probably already made this observation yourself. Solid fats like Crisco and butter are saturated while olive oil, canola oil, and peanut oil are unsaturated and are in the liquid state.

Fatty acids can undergo esterification reactions where the carboxyl group of the fatty acid reacts with an alcohol to produce an ester. Hydration of an unsaturated fatty acid will yield the addition of a hydroxyl group in place of the double bond. Hydrogenation of an unsaturated fatty acid will produce a saturated fatty acid. Halogenation of an unsaturated fatty acid yields the halogenated saturated fatty acid. This is a common test for the degree of unsaturation of a fatty acid. Unsaturated fatty acids can also be oxidized to produce the corresponding aldehydes where the double bonds once were. Ketones do not form because the hydrocarbon chain of fatty acids is rarely substituted with alkyl groups.

b. **Triacylglycerols.** Triacyglycerols or triglycerides are compounds containing three fatty acids esterified to glycerol. The two classes of triglycerides are the fats and oils. Fats tend to be solids at room temperature and are derived from saturated fatty acids esterified to glycerol. Oils are liquids at room temperature and are derived from unsaturated fatty acids esterified to glycerol. Oils can become rancid because they are susceptible to oxidation, which results in volatile organic acids and aldehydes. The degree of unsaturation of oils can be determined with a halogen like iodine. Iodine will add across the double bonds by a halogenation reaction. The amount of iodine absorbed by an oil, the iodine number, can be used to determine the number of carbon-carbon double bonds in 100 grams of sample. Triacylglycerols contain ester bonds, which makes them susceptible to base catalyzed hydrolysis or saponification. The saponification of a triacylglycerol will result in three fatty acid salts and glycerol.

Triacylglycerols can store up to 9 Cal of energy per gram as compared to only 4 Cal per gram for carbohydrates. Most of the energy storing triglycerides in the body are synthesized by the body, but a few are not. The essential fatty acids include linoleic and linolenic acids that are used by the body to synthesize other biomolecules must be obtained by a proper diet.

c. **Complex lipids of membranes and blood.** Cell membranes serve as barriers between cells and their surroundings. They are composed primarily of lipids. The most common lipids in cell membranes include the general classes of polar lipids known as phosphoacylglycerols, sphingolipids, and glycolipids. Phosphoacylglycerols are triacylglycerols that contain a phosphate group in between glycerol and an alcohol. Glycerol contains three hydroxyl groups. Two of them form ester linkages with fatty acids and the third binds phosphoric acid through an ester linkage. If an alcohol combines to the phosphate group, the name of the phosphoacylglycerol will be phosphatidyl followed by the name of the alcohol. Sphingolipids contain the base sphingosine, a fatty acid, and one or more other molecules. Sphingosine can bind molecules through an amine functional group or a hydroxyl functional group. The fatty acid of a sphingolipid bonds to the amine functional group to make an amide linkage. The other molecules that bond to sphingosine form either ester linkages or glycosidic linkages depending on whether a sphingomyelin or a cerebroside is formed. Glycolipids are any lipids that contain a carbohydrate.

d. **Simple lipids.** Simple lipids are nonsaponifiable lipids that consist of steroids, prostaglandins and leukotrienes. All steroids consist of a common framework known as the steroid nucleus that is comprised of four fused rings. Cholesterol is the most abundant steroid that is both synthesized by the body and ingested as part of a normal diet. Cholesterol serves as a building block for the synthesis of many of the sex hormones including testosterone, estrogen, and progesterone. Bile salts are steroids that contain a carboxylate ion that increases the solubility of these lipids in water. Bile salts play a key role in the absorption of fats in the intestines. Prostaglandins and leukotrienes are derivatives of fatty acids. The prostaglandins play key roles in the bodies response to invasion by regulating pain and fever. Leukotrienes are responsible for the body's allergic response by regulating both allergic and inflammation responses.

e. **Membranes and related structures.** Membranes and related structures cover the areas of micelles, lipid bilayers, and the fluid mosaic model. Each of these areas is dependent upon the amphipathic nature of some lipids. The polar head groups of lipids are said to be hydrophobic because they hydrogen bond and are soluble in water. The hydrocarbon tail section of an amphipathic lipid is hydrophobic because it is nonpolar and will therefore avoid water. A micelle is formed when amphipathic lipids are placed in a solution containing water and a layer of grease or oil floating on top. The amphipathic lipids in this case are referred to as soaps or detergents and the aqueous solution with the layer of grease or oil is easily created in your kitchen sink. The grease is nonpolar and will not mix with the water. The addition of amphipathic lipids results in tiny droplets of grease being surrounded by the nonpolar tails of the lipids. The grease is in a nonpolar environment as the nonpolar tails surrounds it. What this does is put the polar head groups out toward the polar aqueous environment. Since the polar heads are hydrophilic, the micelle (lipid/grease complex) becomes soluble in water and you are able to wash the grease away. The bubbles you see when you add the detergent or soap are a result of the micelle formation.

Lipid bilayers are formed much like a micelle except rather than the nonpolar tails surrounding a drop of grease, they stabilize the nonpolar tails of other lipids. Lipids join together in a head to tail to tail to head arrangement and then come around in a big ring like a Certs candy. The nonpolar tails stabilize one another and the aqueous surroundings only come into contact with the polar head groups. The fancy name for the Certs candy arrangement is a liposome.

The fluid mosaic model goes one step further. Take your Certs candy (liposome) model and incorporate steroids, carbohydrates, and proteins. Steroids like cholesterol are nonpolar molecules that will find themselves at home in the hydrophobic interior region of lipid bilayer. The addition of cholesterol creates spaces in the core region that make the membrane move more freely or become more fluid. Carbohydrates are relatively polar molecules because of the hydroxyl groups, so they reside on the exterior of the membrane in the hydrophobic region. Proteins, as you will see in the next chapter, contain both hydrophobic and hydrophilic portions. The hydrophobic portions of a protein will be on the inside of the membrane and the hydrophilic portions will be on the outside of the membrane.

The components of the fluid mosaic model are significant in attempting to understand how cells maintain their identity. Small molecules of carbon dioxide and oxygen will flow into or out of a cell following a concentration gradient by simple diffusion. If the concentration of oxygen is higher outside the cell than inside the cell, oxygen will pass through the membrane from an area of higher concentration to an area of lower concentration. Cells also need access to polar molecules and ions. Small substances like monosaccharides and sodium ions can be transported through the nonpolar interior of a membrane and into a cell through a polar channel in the interior of a protein. This process is called facilitated diffusion and does not require energy because molecules are moving with a concentration gradient. When polar molecules do move against a concentration gradient into a cell, the process is called active transport and it does require energy. Proteins also facilitate active transport.

The four fat-soluble vitamins that you should know are vitamins A, D, E, and K. Vitamin A is important in proper vision, vitamin D is involved in calcium uptake and proper bone mineralization, vitamin E is an antioxidant, and vitamin K is needed for proper blood coagulation.

Common terms you will hear as you get older are LDL and HDL. LDL stands for low-density lipoproteins that transport cholesterol and other lipids from the liver to the cells of the body. If you have too much LDL in your system, the excess cholesterol will deposit on the walls of the arteries resulting in arteriosclerosis. LDL has earned the name "bad cholesterol". HDL is the "good cholesterol" because it functions to transport cholesterol from the body cells back to the liver. Thus HDL helps to clean the excess cholesterol from the arteries and LDL brings more cholesterol into the system.

Solutions to Odd Numbered Problems:

1. A common characteristic of all lipids is that they are generally soluble in nonpolar solvents and insoluble in water.

3. The common complex lipids introduced in this chapter include triacylglycerols, glycolipids, sphingolipids, and phospholipids.

5. Chloroform (a), ethanol (b), and toluene (d) are nonpolar or relatively nonpolar solvents that would dissolve lipids from biological tissues.

7. Fatty acids are different from other carboxylic acids because they have long hydrocarbon chains on the order ot 10 to 24 carbon atoms in length.

9. Saturated fatty acids do not contain any carbon-carbon double bonds. Monounsaturated fatty acids contain one carbon-carbon double bond. Polyunsaturated fatty acids contain more than one carbon-carbon double bond.

11. A cis- double bond will cause a kink or bend in the long chain portion of a fatty acid.

13. The common unsaturated fatty acids that contain one, two, and three carbon-carbon double bonds are oleic acid, linoleic acid, and linolenic acid, respectively.

a.) Oleic acid $$CH_3(CH_2)_7-\overset{H}{\overset{|}{C}}=\overset{H}{\overset{|}{C}}-(CH_2)_7COOH$$

b.) Linoleic acid $$CH_3(CH_2)_4-\overset{H}{\overset{|}{C}}=\overset{H}{\overset{|}{C}}-\underset{H_2}{C}-\overset{H}{\overset{|}{C}}=\overset{H}{\overset{|}{C}}-(CH_2)_7COOH$$

c.) Linolenic acid $$CH_3CH_2-\overset{H}{\overset{|}{C}}=\overset{H}{\overset{|}{C}}-\underset{H_2}{C}-\overset{H}{\overset{|}{C}}=\overset{H}{\overset{|}{C}}-\underset{H_2}{C}-\overset{H}{\overset{|}{C}}=\overset{H}{\overset{|}{C}}-(CH_2)_7COOH$$

15.

Palmitic acid
16:0

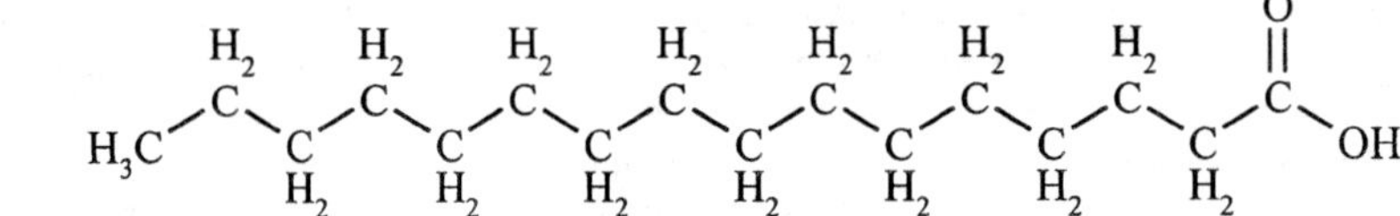

17. a. Oleic acid (18:1) + 1 $H_2 \rightarrow$ Stearic acid (18:0)
b. Palmitic acid only contains a double bond between carbon and oxygen of the carbonyl group. Since the author did not address the hydrogenation of carboxylic acids, it may be assumed that it does not occur.
c. Linolenic acid (18:3) + 3 $H_2 \rightarrow$ Stearic acid (18:0)

19. The ester derived from linoleic acid contains two carbon-carbon double bonds. If either one or both of these bonds were to be switched to the trans- orientation, a different isomer would be formed.

21. Fats and oils are made of three fatty acids esterified to glycerol.

23. Partial hydrogenation of an oil means the some but not all of the carbon-carbon double bonds have been hydrogenated to yield carbon-carbon single bonds in their place.

25. Rancidity is caused by the oxidation of unsaturated hydrocarbon chains in fats or oils. To prevent rancidity, antioxidants can be added to the triglycerides, or they can be partially hydrogenated.

27. The saponification of a triacylglycerol will yield glycerol and three fatty acid salts (soap).

29. If you hydrogenate an oil, you will get a hard solid reminiscent of a fat.

31. Essential fatty acids are compounds that your body does not synthesize, but they are needed by your body for the synthesis of other biomolecules. The essential fatty acids are linoleic and linolenic acids.

33.

Phosphatidic acid

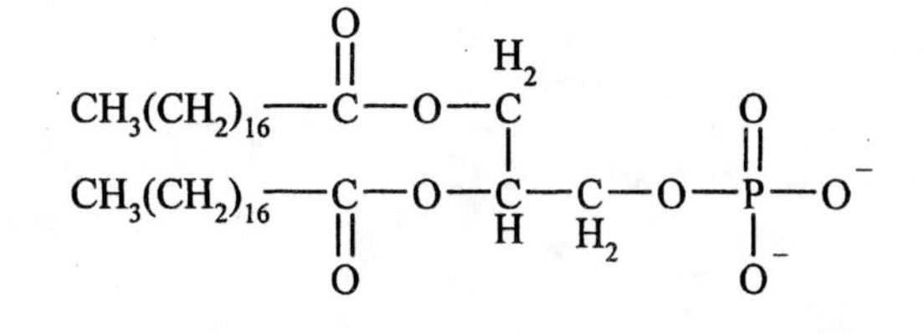

35. The phospholipid lecithin contains phosphatidic acid and the alcohol choline bonded to the phosphate group by an ester bond. The components of phosphatidic acid are glycerol, two fatty acids, and a phosphate group.

37. The difference between a triacylglycerol and a phosphoacyl glycerol is that the triacyglycerol has three fatty acids and the phosphoacylglycerol only contains two fatty acids and a phosphate group.

39. Cephalin and lecithin differ in the alcohol that is attached to the phosphate group. Cephalin has ethanolamine as the alcohol and lecithin has choline as the alcohol.

41. The complex lipid shown has a glycerol, two fatty acids, a phosphate group, and the alcohol ethanolamine. The complex lipid is a phosphoacylglycerol or more specifically, a phosphatidyl ethanolamine.

43. The steroid most abundant in animal membranes is cholesterol.

45. Cholesterol is transported by the blood. When there is an excess of cholesterol in the blood, it can deposit on the linings of the arteries. When sufficient amounts of cholesterol deposit on the arterial lining, the condition is referred to as atherosclerosis.

47. Anabolic steroids increase the body's reactions to synthesis and growth. An athlete would want to take anabolic steroids to increase their normal muscle growth and perform above their natural ability.

49. Hydrocortisone is responsible for the body's anti-inflammatory response.

51. A prostaglandin is a simple lipid that was synthesized from a fatty acid and is responsible for the regulation of body function through pain response and fever response.

53. Cholesterol (a)

55.

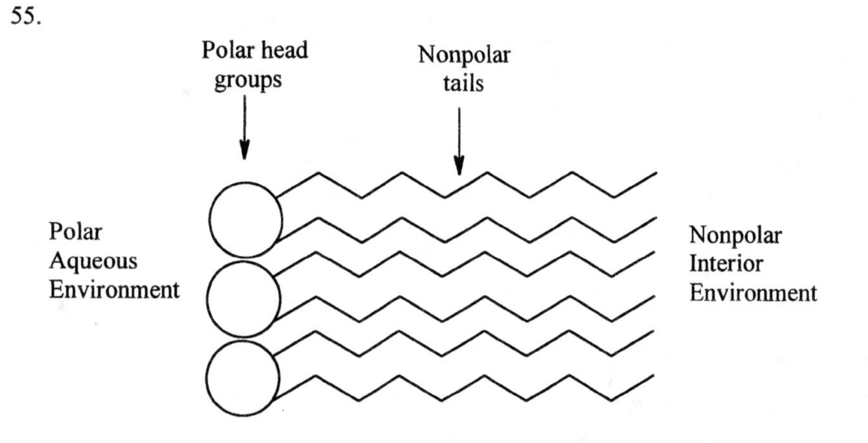

57. Hydrophobic interactions

59. The aqueous core of a liposome could be a good place to put a pharmaceutical drug for improved delivery.

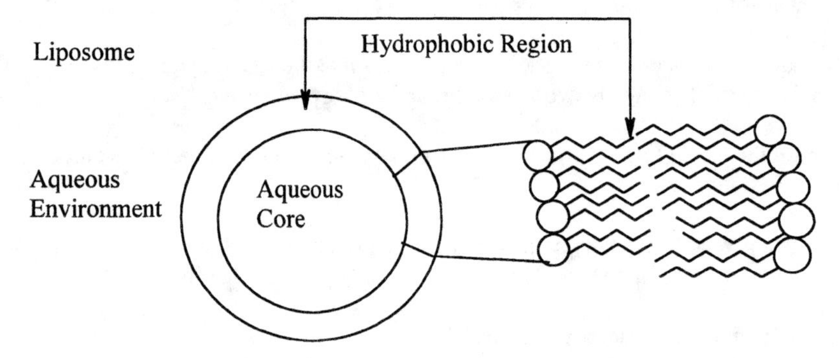

61. The fluid mosaic model is the current model for cellular membranes that has a lipid bilayer with proteins and cholesterol in the interior hydrophobic portion, and carbohydrates and hydrophilic protein portions on the exterior. Proteins are free to move in the bilayer and form a patchwork pattern giving both the qualities of fluidity and a mosaic.

63. Phospholipids (c)

65. a. Lipids are biological molecules that are insoluble in water but soluble in organic solvents.
b. Triacylglycerols are compounds containing three fatty acids esterified to glycerol.
c. Unsaturated fatty acids are long chain carboxylic acids that contain one or more carbon-carbon double bonds.
d. A fat is a solid triacyglycerol that contains few if any degrees of unsaturation.
e. A phosphoacylglycerol is a molecule of phosphatidic acid with an alcohol connected to the phosphate group through an ester linkage.
f. Facilitated transport is the protein-assisted movement of ions or polar molecules through a membrane, along a concentration gradient, without the expenditure of energy.
g. Phosphatidyl choline is a phosphoacylglycerol where the alcohol bonded to the phosphate group is choline.
h. A glycolipid is a lipid that contains a carbohydrate as part of its structure.
i. Prostaglandin is a type of lipid synthesized from a polyunsaturated fatty acid that is thought to be responsible for the body's regulatory properties of fever and pain.
j. Steroids are nonsaponifiable lipids that contain four rings fused in a pattern termed the steroid nulceus.
k. Estrogen is a steroid that is the common female sex hormone containing an aromatic ring as part of the steroid nucleus.
l. Testosterone is common male sex hormone that is a steroid.
m. Bile salts are amphipathic steroids that emulsify (breakdown) dietary lipids.

n. Cholesterol is the most common amphipathic steroid found in animals that keeps membranes fluid and serves as a starting place for the synthesis of many other steroids.
o. Hydrocortisone is an adrenocorticoid hormone also known as cortisol that is a powerful antiinflammatory agent.

67. The individual has 20% of their total body mass of 50 kg as fat. Twenty percent of fifty kilograms is ten kilograms or 10,000 grams. Every gram of fat will supply nine Calories of energy (p. 402). Multiplying the terms together gives a total energy storage of 90,000 Calories. If the individual burns 2,000 calories a day, it will take 45 days before they have used up all of their fat reserves.

Sample Test:

1. Lipids are defined as biological molecules that are
   a. Soluble in water, but insoluble in nonpolar solvents.
   b. Insoluble in both water and nonpolar solvents.
   c. Soluble in nonpolar solvents but insoluble in water.
   d. Soluble in both water and nonpolar solvents.

2. In which of the following pairs of fatty acids does the first listed acid have a melting point that is higher than the second listed fatty acid?
   a. 16:1 and 16:0
   b. 20:0 and 18:0
   c. 18:3 and 18:1
   d. 18:2 and 20:0

3. Fats and oils are referred to as all of the following except
   a. Triacylglycerols
   b. Triglycerides
   c. Glycerol triesters
   d. Fatty acid salts.

4. Unsaturated fatty acids are found in which of the following?
   a. Both fats and oils b. Oils but not fats c. Fats but not oils d. Neither fats nor oils

5. Hydrolysis of an oil will produce three fatty acids and which of the following?
   a. A long-chain alcohol b. Glycerol c. Choline d. Ethanolamine

6. The saponification of a fat will produce which of the following?
   a. Glycerol and three fatty acids.
   b. Glycerol and three fatty acid salts.
   c. Both fatty acids and fatty acid salts.
   d. Glycerol, fatty acids, and fatty acid salts.

7. Which of the following statements concerning fatty acids is correct?
   a. Fatty acids are commonly encountered as the free fatty acid in nature.
   b. Unsaturated fatty acids almost always have the cis-configuration.
   c. The hydrocarbon chain of a fatty acid must contain at least 20 carbon atoms.
   d. All fatty acids are essential and must be obtained from the diet.

8. All phosphoacylglycerols contain the following components esterified to glycerol.
   a. One fatty acid and two phosphoric acid molecules.
   b. Two fatty acid and one phosphoric acid molecules.
   c. Three phosphoric acid molecules.
   d. One fatty acid, one phosphoric acid, and one ethanolamine.

9. Which of the following lipids are saponifiable?
   a. Prostaglandins b. Leukotrienes c. Sphingolipids d. Androgens

10. Sphingolipids contain sphingosine bonded to which of the following components?
    a. Two fatty acids participating in ester linkages.
    b. Two fatty acids participating in amide linkages.
    c. One fatty acid participating in an ester linkage.
    d. One fatty acid participating in an amide linkage.

11. What is the most abundant steroid in the human body?
    a. Estrogen b. Testosterone c. Progesterone d. Cholesterol

12. All steroids contain a steroid nucleus containing what fused-ring system?
    a. Four six-membered rings.
    b. Four five membered rings.
    c. Three six-membered rings and one five-membered ring.
    d. Two six-membered rings and two five-membered rings.

13. Which of the following lipids is not a steroid?
    a. Hydrocortisone b. Bile salts c. Anabolic substances d. Phosphatidyl ethanolamine

14. The partial hydrogenation of a fat or oil will result in which of the following?
    a. The production of fatty acid salts.
    b. A fat or oil with a lower melting point.
    c. A fat or oil with a higher boiling point.
    d. A fat or oil with fewer carbon-carbon double bonds.

15. All of the following components are common in plasma membranes except
    a. Cholesterol b. Phosphoacylglycerols c. Prostaglandins d. Glycolipids

16. Which of the following statements concerning lipid bilayers is correct?
    a. Both the exterior and interior surface of a lipid bilayer contain groups that are nonpolar.
    b. The interior of a lipid bilayer is polar.
    c. The two layers of molecules in a lipid bilayer are present in a "tail-to-tail" arrangement.
    d. The two layers of molecules in a lipid bilayer are present in a "head-to-tail" arrangement.

17. Which of the following lipids are not amphipathic?
    a. Cholesterol b. Phospholipids c. Bile salts d. They all are.

18. A polar molecule could be passed through the nonpolar membrane medium without the expenditure of energy by what process?
    a. Simple diffusion b. Facilitated diffusion c. Active transport d. Energy is required.

19. Which fat-soluble vitamin plays a role in calcium uptake and bone mineralization?
    a. Vitamin A b. Vitamin D c. Vitamin E d. Vitamin K

20. A cerebroside would fit into what class of lipid?
    a. Sphingolipid b. Glycolipid c. Phospholipid d. Triglyceride

Answers to Sample Test:

1. c
2. b
3. d
4. a
5. b

6. b
7. b
8. b
9. c
10. d
11. d
12. c
13. d
14. d
15. c
16. c
17. d
18. b
19. b
20. a

Chapter 14: Proteins

Chapter Objectives:

1. To recognize and categorize by functional group the twenty common amino acids.
2. To describe and understand peptide bonds and what they do.
3. To recognize the differences between primary, secondary, tertiary, and quaternary protein structure.
4. To become familiar with commonly encountered proteins.

Key Terms: The key terms are listed in the order in which they are encountered in the chapter.

Proteios. The first rank or importance.

Proteins. The most abundant class of biological polymers, in the body, made up of monomer units called amino acids.

Amino acid. Organic compounds that contain both an amino group and a carboxyl group.

α-amino acids. The twenty common amino acids that have a carboxyl group, amino group, hydrogen atom, and side chain bonded to the same carbon atom called the alpha carbon atom.

α-Carbon atom. The alpha carbon atom of an amino acid is a chiral center attached to an amino group, a carboxyl group, a hydrogen atom, and a side chain with the exception of glycine, which has a hydrogen atom as a side chain.

Nonpolar amino acids. Alpha amino acids that contain nonpolar hydrocarbon side chains attached to the alpha carbon atom.

Polar amino acids. Alpha amino acids that contain a polar side chain attached to the alpha carbon atom.

Neutral polar amino acids. Alpha amino acids that contain side chains that are polar but not usually ionic, like alcohols, thiols, phenols, amides, and thioethers.

Acidic amino acids. Alpha amino acids that contain a side chain with a carboxyl functional group.

Basic amino acids. Alpha amino acids that contain a side chain with one or more nitrogen atoms.

Zwitterion. An electrically neutral molecule that contains both a positively charged ion and a negatively charged ion that is water-soluble and has a high melting point.

Amphoteric molecules. Molecules that possess both acidic and basic properties.

Isoelectric point (pI). The pH value at which an amino acid has no net charge.

Essential amino acids. Amino acids that cannot be synthesized by our body but are required for health and well being.

Peptides. Oligomers and polymers of amino acids that have a molar mass that is generally less than 5000 atomic mass units (amu).

Proteins. Large polymers of amino acids that have molar masses greater than 5000 amu.

Peptide bond. The amide bond that forms between the carbonyl carbon of the carboxyl group of one amino acid and the alpha amino group of another amino acid.

Dipeptide. A peptide containing two amino acids joined together by a peptide bond.

Tripeptide. A peptide containing three amino acids joined together by peptide bonds.

Tetrapeptide. A peptide containing four amino acids joined together by peptide bonds.

Oligopeptide. A peptide containing between up to ten amino acids.

Polypeptide. A peptide containing more than ten amino acids.

N-terminal amino acid. The amino acid of a peptide that has the free amino group is always the first amino acid in the chain.

C-terminal amino acid. The amino acid of a peptide that has the free carboxyl group is always the last amino acid in the chain.

Oxytocin. An oligopeptide containing nine amino acids (nonapeptide) that is responsible for contraction of uterine and smooth muscles.

Vasopressin. An oligopeptide containing nine amino acids (nonapeptide) that is responsible for water retention and blood pressure regulation.

Endorphins. Oligopeptides that are involved in pain and pleasure sensations.

Enkephalins. Oligopeptides that are involved in pain and pleasure sensations.

Native conformation. The conformation of an active protein that is frequently its lowest energy most stable form.

Denaturation. A change in the native conformation of a protein that results in the loss of activity.

Primary structure. The amino acid sequence of a protein that directs the unique shape or conformation of the protein.

Secondary structure. The hydrogen bonding interactions that occur between the atoms of peptide bonds that constitute the backbone of a protein.

α-helix. The helical structure in proteins that is maintained by hydrogen bonds between the amide proton of one amino acid and the carbonyl oxygen atom of another amino acid four residues further down the chain.

β-pleated sheet. The structural characteristic that forms as a result of hydrogen bonding between atoms of the peptide bond that are in distant portions of the same chain or in different polypeptide chains.

Fibrous proteins. Water-insoluble proteins that are rope-like in structure due to their either an α-helical or β-pleated sheet secondary structure.

Globular proteins. Proteins that are compact or roughly spherical as a result of tertiary structure.

Tertiary structure. The hydrogen bonding, hydrophobic, ionic, and disulfide bonding interactions of the side chain atoms that stabilize a proteins structure.

Disulfide bridge. The sulfur-sulfur bond that forms from the oxidation of the thiol functional groups present in cysteine residues.

Quaternary structure. The arrangement of polypeptide chains for a protein that requires more than one polypeptide chain for activity.

Oligomeric protein. A protein that contains up to ten polypeptide subunits.

Subunit. A polypeptide chain or monomer in an oligomeric protein.

Hemoglobin. A protein consisting of four subunits that is responsible for the transport of oxygen in the blood of humans.

Simple proteins. Proteins that contain only amino acids.

Conjugated proteins. A protein that contains amino acids and one or more additional groups called prosthetic groups.

Prosthetic group. A nonamino acid portion of a protein that is necessary for the protein to function.

Heme. The prosthetic group in hemoglobin that is necessary for hemoglobin to bind oxygen.

Antibodies. Proteins that function to protect the body from the invasion of foreign particles called antigens.

Antigen. Foreign particles such as viruses and bacteria that can invade the body and stimulate the production of antibodies.

Autoimmune disease. A condition that occurs when antibodies attempt to damage and destroy the body's own natural components.

Electrophoresis. A technique that uses an applied electric field to cause charged particles like proteins to move at a rate that is characteristic of their molecular size and overall charge.

Chapter 14 Objective Details:

a. Amino acids. Amino acids are the building blocks for proteins. The twenty common amino acids all have the same basic structure. The alpha carbon atom is at the center of every amino acid. Recall from Chapter 11 (Section 11.4), on carboxylic acids, the carbon atom adjacent to the carboxyl functional group was termed the alpha carbon atom. This nomenclature is the commonly accepted way to name the amino acids. Bonded to the alpha carbon atom are the carboxyl group, an amino group, a hydrogen atom (alpha hydrogen), and a side chain. The side chains of the amino acids contain all of the major organic functional groups that you have already learned about.

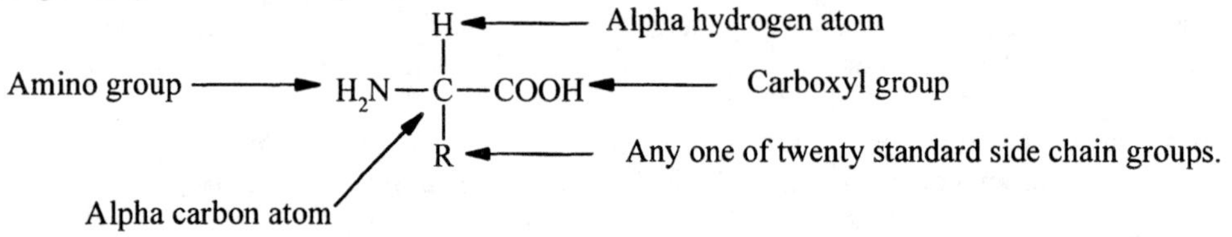

H ← Alpha hydrogen atom

Amino group → $H_2N-C-COOH$ ← Carboxyl group

R ← Any one of twenty standard side chain groups.

Alpha carbon atom

All of the nonpolar alkyl and aryl side chains are classified as nonpolar amino acids. If the side is polar, but not able to be ionized, then the side chain is considered polar neutral. Examples of functional groups that fit into this category are the alcohols, thiols, phenols, thioethers, and amides. If the side chain contains a carboxyl functional group, it is in the category of polar acidic amino acids. Some side chains contain primary, secondary, or tertiary amines that are alkaline, so these amino acids are called the polar basic amino acids.

Amino acids by themselves are chiral with the exception of glycine, which contains a hydrogen atom as a side chain. Amino acids can undergo intramolecular acid-base reactions where the carboxyl group is deprotonated by the amino group. The resulting amino acid contains a negative charge on the carboxyl group and a positive charge on the amino group. A molecule with two equal and opposite charges on different atoms is called a zwitterion. The intramolecular acid-base reaction will occur for amphoteric molecules.

b. The peptide bond. A peptide bond is the amide bond that forms between the carbonyl carbon of the carboxyl group of one amino acid and the alpha amino group of another amino acid. Peptide bonds are what join all of the amino acids in peptides and proteins. The formation of a peptide bond between two amino acids leaves one of the amino acids with a free amino group and the other with a free carboxyl group. The amino acid with the free amino group is called the N-terminal amino acid. In peptide or protein nomenclature, this amino acid is always named first. All of the amino acids are then named in order until the amino acid with the free carboxyl group is encountered. This amino acid is the C-terminal amino acid. Peptides of up to ten amino acids can be referred to as oligopeptides. As the size of the peptide increases beyond ten amino acids, the term polypeptide is used. If the size of the molecule contains greater than 5000 atomic mass units, it is referred to as a protein.

c. Protein structure. Protein structure consists of primary, secondary, tertiary, and quaternary structure. The simplest way to think of each level of protein structure is to break it down to the basics. An overview is as follows:

1.) Primary structure: the amino acid sequence.
2.) Secondary structure: hydrogen bonding interactions of the backbone atoms.
3.) Tertiary structure: hydrogen bonding, hydrophobic, ionic, disulfide bonding interactions of side chain atoms.
4.) Quaternary structure: Interactions between polypeptide chains.

1.) The primary structure of a protein is the amino acid sequence. Every protein has a different sequence and it is this sequence that determines the conformation of the final protein.
2.) The secondary structure of a protein involves the hydrogen bonding interactions between the amide proton of one amino acid and the carbonyl oxygen of another amino acid. The hydrogen bonds exist between atoms of the peptide bonds that hold the protein together. The two types of backbone bonding patterns are the α-helix and the β-pleated sheet. The α-helix involves the hydrogen bonding interactions between amide proton of one amino acid and the carbonyl oxygen of another amino acid four residues away. The amino acids are part of the same chain and form a helical structure. The β-pleated sheet involves hydrogen bonding interactions between the amide hydrogen of one amino acid and the carbonyl oxygen atom of another amino acid that are either distant portions of the same chain or on different chains. If you can imagine a sheet that is either folded back on itself or that is on top of another sheet, this is the same idea.
3.) The tertiary structure of a protein involves the interactions of the side chain atoms. In Section 14.1, you observed that there are four types of amino acids that differ only in the functionality of their side chains. When those four classes of amino acid side chains interact, they can do so in any one of four different ways. The tertiary structure of proteins involves hydrogen bonding, hydrophobic, ionic, and disulfide bonding. Hydrogen bonding interactions are weak dipole-dipole interactions between a hydrogen atom bonded to oxygen, nitrogen, or fluorine that is attracted to an electronegative atom of oxygen, nitrogen, or fluorine. Hydrophobic interactions are the London dispersion forces that attract nonpolar alkyl substituents. Ionic interactions are the forces of attraction that exist between oppositely charged side chains. Disulfide bonds are covalent bonds that connect the two sulfur atoms of two different cysteine residues.
4.) Quaternary structure exists when two or more polypeptides are necessary for the activity of a protein. These polypeptides can be held together by the same interactions that are involved in tertiary structure. Hemoglobin is the common example of a protein that has quaternary structure.

Simple proteins contain amino acids where conjugated proteins contain amino acids and a prosthetic group. Prosthetic groups are nonamino acid groups that are necessary for the proper functioning of the protein. You will learn more about prosthetic groups when you cover minerals. Many metal ions including zinc, magnesium, manganese, iron, and calcium are essential for the proper functioning of many proteins.

d. Some functions and properties of proteins. Proteins can function as enzymes, structural components of cells, transporters of molecules across membranes, protectors of tissue, motion of tissue, energy storage, and regulation of cell and body activities. An example of protective proteins is the antibodies that your body produces to destroy bacterial or viral particles called antibodies. The activity of a protein is typically controlled by the ability of the protein to maintain its native conformation in the body. Thus, the solubility of the protein is critical. Protein solubility is maintained by the overall influence of the charges on the side chains exposed to the exterior polar environment. Acidic residues are typically deprotonated and have a negative charge while alkaline amino acids have a positive charge at neutral pH. These charges contribute to protein solubility and are the reason why proteins will denature when the pH of their surroundings increases or decreases. Other factors that will cause proteins to denature are heat, organic solvents, ionic solutions, and the presence of heavy metal ions. A mixture of proteins can be separated and analyzed by electrophoresis. Electrophoresis is a method used to separate proteins based on size and charge. By running an electric current through a polymer gel, proteins will move at different rates allowing for their isolation, purification, and identification.

Solutions to Odd Numbered Problems:

1.

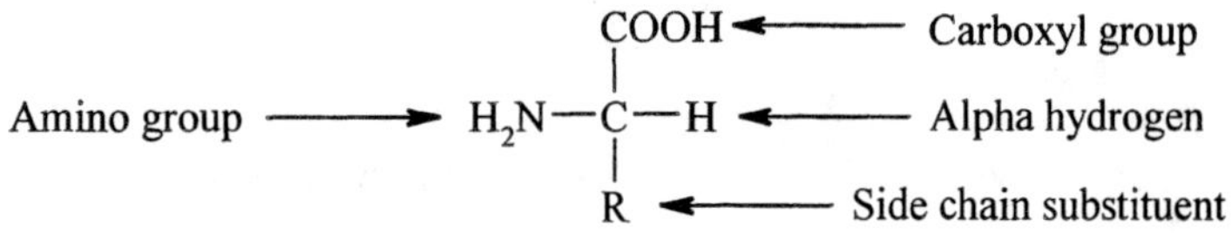

COOH ← Carboxyl group

Amino group → $H_2N—C—H$ ← Alpha hydrogen

R ← Side chain substituent

3. (b) All of the common amino acids are in the L-stereochemical family.

5. The structure of L-valine in its zwitterion form is shown below.

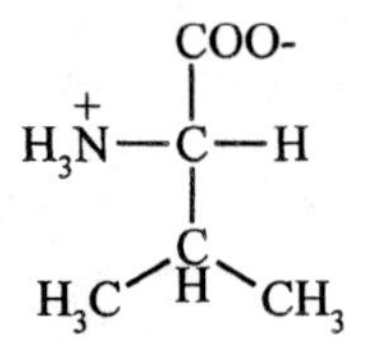

7. Threonine contains two chiral centers. The alpha carbon is the chiral center compared to the glyceraldehydes to determine the stereochemical family.

9. a. Phe = phenylalanine
   b. Cys = Cysteine
   c. Glu = Glutamic acid
   d. Lys = Lysine

11. Nonpolar amino acids contain only nonpolar hydrocarbon side chains that are either alkyl or aryl. The polar neutral, acidic, or basic amino acids all contain atoms in their side chains other than carbon or hydrogen.

13. Proline is unique because the end of its side chain is bonded to the alpha amino group making a ring.

15. The side chains for aspartic acid and lysine will often be charged when these amino acids are in zwitterionic form, but their charge is not involved in the definition of the zwitterion.

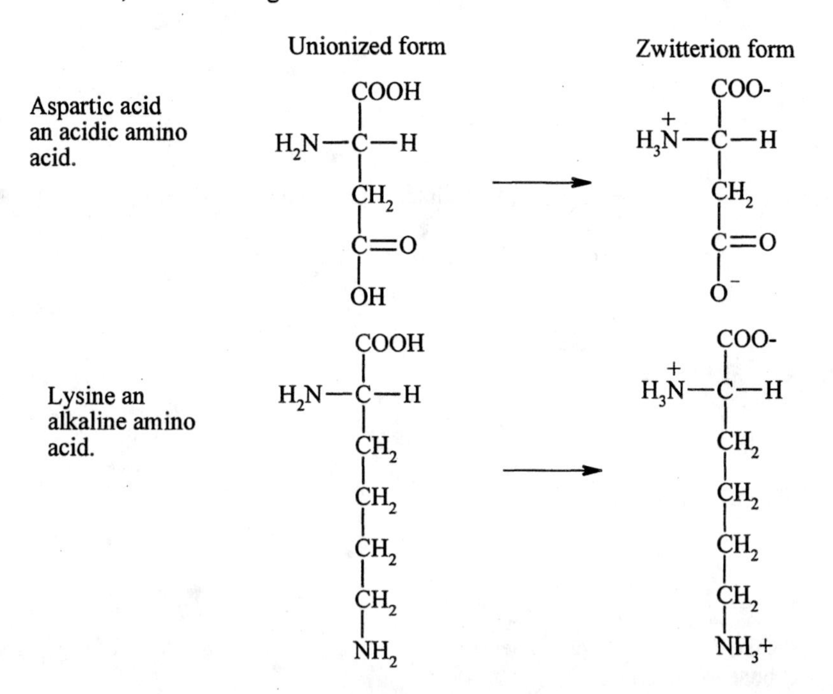

17. There is not a pH value where all of the functional groups of an alpha amino acid are un-ionized. At low pH values, the amino group will have a positive charge, neutral pH values the amino group will be positive and the carboxyl group will be negative, and at high values of pH, the carboxyl group will have a negative charge.

19.

Aspartic acid pH = 1 → pH = 7 → pH = 13

pH = 1: $\overset{+}{H_3N}$—CH(COOH)—CH_2—C(=O)—OH

pH = 7: $\overset{+}{H_3N}$—CH(COO-)—CH_2—C(=O)—O^-

pH = 13: H_2N—CH(COO-)—CH_2—C(=O)—O^-

21. The essential amino acids are arginine, histidine, isoleucine, leucine, lysine, methionine, phenylalanine, threonine, tryptophan, and valine.

23. A peptide bond is the amide bond that forms between the carbonyl carbon of the carboxyl group of one amino acid and the alpha amino group of another amino acid.

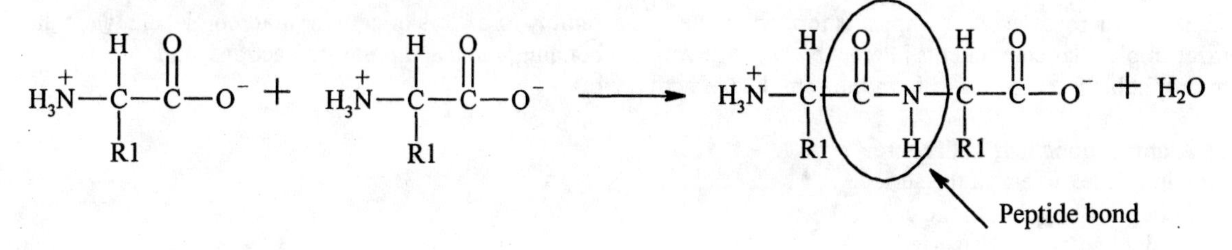

25.

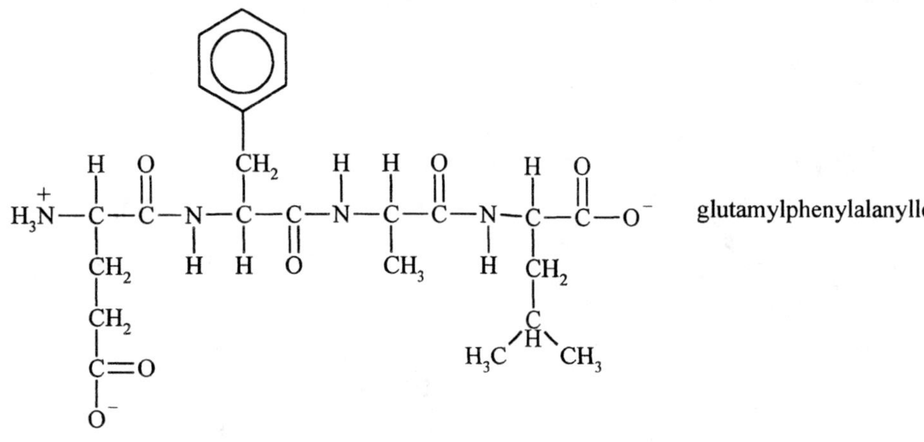

glutamylphenylalanylleucine

27. Alanlyglycylvaline

29. a. Hemoglobin is a protein. It is too big to be a peptide.
    b. Casein is a protein.
    c. Myoglobin is a protein.
    d. Vasopressin is a peptide.

31. The secondary structure of a protein is the hydrogen bonding interactions that occur between the atoms of peptide bonds that constitute the backbone of the protein.

33. The tertiary structure of a protein deals with the interactions of the side chain atoms that create the compact shape of a protein's structure.

35. The forces that are responsible for protein primary structure are the peptide bonds that hold the amino acids of the protein sequence together.

37. The forces that are responsible for protein tertiary structure are hydrogen bonding, hydrophobic, ionic, and disulfide bonding interactions.

39. (b) Secondary structure

41. (b) Secondary structure

43. (c) Hydrophobic interactions

45. A disulfide bridge is the product of the oxidation of two thiols, which are the functional groups present in cysteine residues that result in the formation of a sulfur-sulfur bond. The sulfur-sulfur bond is a covalent bond that maintains the tertiary structure of a protein and where appropriate, the quaternary structure of an oligomeric protein.

47. A stable macromolecular complex is formed by the self-assembly of a virus or cellular macromolecule when the macromolecular components interact by way of hydrogen bonding, ionic and polar interactions, and hydrophobic interactions to form the complete or active virus.

49. a. Alanine; nonpolar; in the core
    b. Glutamate; polar; on the surface
    c. Leucine; nonpolar; in the core
    d. Lysine; polar; on the surface
    e. Threonine; polar; on the surface
    f. Valine; nonpolar; in the core

g. Phenylalanine; nonpolar; in the core

51. The amino acids that could form ionic interactions when in a protein are glutamate and lysine.

53. The amino acids that could form hydrophobic interactions when in a protein are the nonpolar amino acids alanine, leucine, valine, and phenylalanine.

55. Lysine, arginine, and histidine would have a positive charge and glutamate and aspartate would have a negative charge at a pH of 7.

57. a. Enzymes are biological catalysts that lower the activation energy of biological reactions.
    b. Antibodies destroy antigens, which are viral or bacterial particles.
    c. Serum albumin transports fatty acids in the blood.
    d. Hemoglobin transports oxygen in the blood.
    e. Myoglobin stores oxygen in the muscles.
    f. Lysozyme destroys cell walls of some bacteria.
    g. Insulin is a regulatory protein involved in the use of nutrients by the body.
    h. Collagen is a fibrous protein that provides the structural component of connective tissue.

59. A protein with an isoelectric point of 6.13 will become negatively charged at a pH of 7 and even more negatively charged at a pH of 9. When the pH is lowered to 5.5, the protein will have a more positive charge.

61. The isoelectric point of a protein is the pH at which the protein has no net charge.

63. (b) Ionic interactions

65. a. α-amino acids have a carboxyl group, an amino group, a hydrogen atom, and a side chain bonded to the same carbon atom.
    b. L-amino acids are the common twenty amino acids that have the amino group on the left of the alpha carbon atom when the carboxyl group is drawn up and the side chain is drawn down.
    c. A zwitterion is an electrically neutral molecule that contains both a positively charged ion and a negatively charged ion.
    d. Amphoteric molecules possess both acidic and basic properties.
    e. The isoelectric point of an amino acid or protein is the pH value at which the molecule has no net charge.
    f. A dipeptide is two amino acids joined by a peptide bond.
    g. A peptide bond is the amide bond that forms between the carbonyl carbon of the carboxyl group of one amino acid and the alpha amino group of another amino acid.
    h. A protein is a polymer of amino acid monomer subunits that has a mass in excess of 5,000 amu.
    i. The primary structure of a protein is the sequence of amino acids.
    j. The secondary structure of a protein is the alpha helices and beta pleated sheets that are held together by hydrogen bonding interactions that form between the amide proton of one amino acid and the carbonyl oxygen of another amino acid in the backbone of the polymer chain.
    k. The tertiary structure of a protein is the overall shape of a protein stabilized by hydrogen bonding, hydrophobic, ionic, and disulfide bonding interactions of the side chain atoms.
    l. Quaternary structure of a protein is the arrangement of polypeptide chains in an oligomeric protein that requires more than one polypeptide chain for activity.
    m. Hydrophobic interactions are the London dispersive forces that cause nonpolar molecules to be attracted to one another.
    n. A disulfide bridge is a covalent bond between two sulfur atoms that formed as the result of the oxidation of the thiols of two cysteine residues.
    o. Denaturation is a change in the native conformation of a protein that results in a loss in activity.

67. a. Sickle cell anemia is caused by a change in the identity of the sixth amino acid of hemoglobin from a glutamic acid to a valine.
    b. Glutamic acid is an acidic residue where valine is nonpolar. The difference in polar properties of these two amino acids is enough to make hemoglobin insoluble in the blood when valine is present. The decreased

solubility of the valine containing hemoglobin alters the shape of the cell and decreases its ability to carry oxygen.
c. Two treatments for sickle cell anemia include blood transfusion and use of the polymerase chain reaction to produce sufficient quantities of normal hemoglobin for injection.

Sample Test:

1. Proteins are all of the following except
   a. The most abundant biological molecule in the body.
   b. They all contain the elements C, H, O, and N.
   c. They can have more than one polypeptide chain present.
   d. Once formed, their native conformation cannot be disrupted.

2. The standard amino acids are all of the following except
   a. Twenty in number.
   b. Alpha amino acids.
   c. They possess at least one chiral center.
   d. They are left-handed except one.

3. The difference between each of the standard amino acids is
   a. The location of the amino group.
   b. The location of the carboxyl group.
   c. The location of the alpha carbon atom.
   d. The composition of the side chain.

4. A reaction between what two functional groups results in a peptide bond?
   a. Two amino groups.
   b. Two carboxyl groups.
   c. An amino group and a carboxyl group.
   d. A carboxyl group and a hydroxyl group.

5. How many peptide bonds are present in glycylphenylalanlyvalylserine?
   a. 2 b. 3 c. 4 d. 5

6. What is the N-terminal amino acid in the tetrapeptide Tyr-Cys-Trp-Val?
   a. Tryptophan b. Tyrosine c. Valine d. Cysteine

7. Which of the following levels of protein structure involves the interactions of the side chain atoms?
   a. Primary structure.
   b. Secondary structure.
   c. Tertiary structure.
   d. Quaternary structure.

8. What proteins have quaternary structure?
   a. Proteins that have polar amino acid residues.
   b. Proteins that transport oxygen.
   c. Proteins that contain more than one subunit.
   d. Proteins that are held together by hydrogen bonding interactions.

9. Which of the following could not be a prosthetic group of a conjugated protein?
   a. A metal ion. b. A heme group. c. Alanine d. Vitamin K

10. Which of the following is not a protein denaturing agent?
    a. An aqueous solution.
    b. An ionic solution.
    c. An organic solution.

d. Heat

11. Which of the following elements is present in some of the amino acids but not all of them?
    a. Oxygen b. Carbon c. Sulfur d. Nitrogen

12. What will be the charge on the alkaline amino acid Lysine at a pH of 7?
    a. Lysine will have no net charge and will exist as the zwitterion.
    b. Lysine is a base so it will have a negative charge.
    c. Lysine is a base so it will have a positive charge.
    d. Lysine has two amino groups so it will have a positive two overall charge.

13. Which of the following interactions involves the sharing of a pair of electrons?
    a. Hydrogen bonding
    b. Hydrophobic interactions
    c. Disulfide bridges
    d. Ionic interactions.

14. Which of the following levels of protein structure is not disrupted when protein denaturation occurs?
    a. Primary structure.
    b. Secondary structure.
    c. Tertiary structure.
    d. Quaternary structure.

15. During the acid catalyzed hydrolysis of proteins that occurs in your stomach, what level of protein structure is not disrupted?
    a. Primary structure.
    b. Secondary structure.
    c. Tertiary structure.
    d. All levels of protein structure are distrupted.

16. What is the function of the proteins referred to as antibodies?
    a. They are involved in regulation of growth and metabolism.
    b. They react with foreign bodies and aid in their removal.
    c. They store oxygen in muscles.
    d. They synthesize proteins.

17. Which of the following molecules is not an oligopeptide?
    a. Oxytocin b. Vasopressin c. Endorphins d. Morphine

18. How does electrophoresis separate proteins?
    a. By function b. By structure c. By charge d. By reactivity

19. Based on the charge of the following amino acid, what is the pH of the solution?

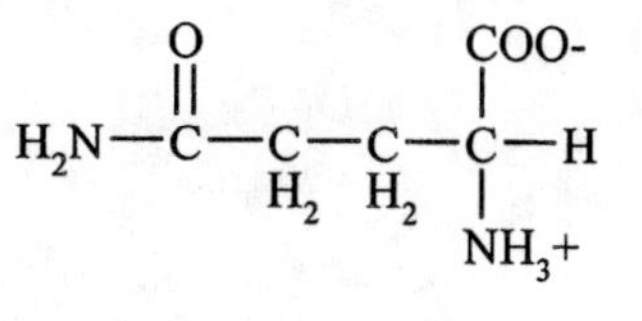

```
        O              COO-
        ||             |
  H2N—C—C—C—C—H
         H2  H2  |
                 NH3+
```

    a. Low pH b. High pH c. Neutral pH d. Not enough information.

20. Sickle cell anemia is the result of a glutamic acid residue being replaced by a valine residue. This change in primary structure will affect what other levels of hemoglobin structure?
    a. Secondary structure
    b. Tertiary structure

c. Quaternary structure
d. All of the above

Answers to Sample Test:

1. d
2. c
3. d
4. c
5. c
6. b
7. c
8. c
9. c
10. a
11. c
12. c
13. c
14. a
15. d
16. b
17. d
18. c
19. c
20. d

Chapter 15: Enzymes

Chapter Objectives:

1. To understand the nomenclature of enzymes.
2. To describe the interactions between enzymes and substrates.
3. To understand the factors involved in enzyme activity.
4. To explain the factors effecting the rates of enzymatic reactions.
5. To differentiate between the types of enzyme regulation.

Key Terms: The key terms are listed in the order in which they are encountered in the chapter.

Enzymes. Protein catalysts that speed up the chemical reactions that occur in the body at the modest temperatures and mild conditions existing in cells.

Substrates. Molecules that react in the enzymatic reactions of biological systems.

Oxidoreductase. One of the six major classes of enzymes that catalyzes redox (oxidation-reduction) reactions.

Transferases. One of the six major classes of enzymes that catalyzes the transfer of one or more atoms from one substance to another.

Hydrolases. One of the six major classes of enzymes that catalyzes hydrolytic cleavage (hydrolysis) or the reverse (dehydration).

Lyases. One of the six major classes of enzymes that catalyzes the addition of a group to a double bond or the removal of a group from a double bond in a manner that does not involve oxidation-reduction or hydrolysis.

Isomerases. One of the six major classes of enzymes that catalyzes intramolecular rearrangements (isomerizations).

Ligases. One of the six major classes of enzymes that catalyzes the bonding together of two substrate molecules with the expenditure of energy.

Cofactor. The organic molecules and inorganic metal ions that serve as prosthetic groups for enzymes.

Coenzyme. The organic molecules that serve as prosthetic groups for enzymes.

Holoenzyme. A conjugated protein; an enzyme coupled with its cofactor.

Apoenzyme. A simple protein; an enzyme lacking its required cofactor.

Active site. The region of an enzyme where the substrate binds and the conversion to product occurs.

Specificity. Enzymes are exposed to many different compounds within a cell, yet they only bind to one or a few of them.

Lock-and-key model. This model views the substrate and active site of an enzyme as being complementary prior to binding.

Induced-fit model. This model views the substrate and active site of an enzyme as having to change in shape in order to accommodate one another prior to catalysis.

Proximity effects. One way that enzymes catalyze reactions is to bring substrate molecules close enough to one another in the active site for a chemical reaction to occur.

Orientation effects. The active site of an enzyme will bind a substrate specifically such that the reactive end of the molecule is exposed to the reactive end of another substrate.

Acid-base catalysis. The active site of an enzyme can contain amino acids that are both close and properly oriented to quickly give up or accept hydrogen ions from the substrate to facilitate the reaction.

Strain. The binding of a substrate to the active site of an enzyme can turn the substrate in to a strained, highly reactive species that will require much less energy than normal to react with another substrate.

Reaction rate. The speed at which a chemical reaction occurs under a set of conditions.

V_{max}. The maximal value of an enzymatic reaction under saturating conditions of substrate.

Homeothermic. The term used to describe an organism that maintains a set body temperature.

Hypothermia. The condition that results when the body temperature drops well below normal.

Fever. The condition that results when the body temperature rises well above normal (hyperthermia).

Inhibitors. Substances that bind to enzymes and alter their catalytic activity.

Irreversible inhibitor. A substance that permanently binds to an enzyme, leaving the enzyme with little or no activity.

Reversible inhibitors. Substances that bind to an enzyme molecule to reduce or eliminate enzyme activity, but only temporarily.

Competitive inhibitor. A substance resembling the substrate that binds to the active site of an enzyme and blocks the substrate from binding, thus inhibiting the catalytic activity of the enzyme.

Noncompetitive inhibitors. Substances that bind to an enzyme in a place other than the active site and cause a conformational change in the enzyme that reduces or eliminates the enzyme's ability to catalyze a reaction.

Proenzyme. A precursor to an enzyme that has no catalytic activity.

Zymogens. Proenzymes of protein-cleaving enzymes that are recognized by the suffix –ogen or the prefix pro-.

Trypsinogen. The inactive precursor (zymogen) of the digestive enzyme trypsin, which is activated upon entry into the small intestine.

Covalent modification. Covalent attachment of a group to an enzyme that results in the reversible increase or decrease of enzyme activity.

Glycogen phosphorylase. The enzyme responsible for cleaving a glucose molecule from glycogen and attaching a phosphate group to it.

Fibrinogen. The zymogen of the blood-clotting enzyme fibrin.

Fibrin. The enzyme that forms insoluble molecules whose aggregation results in the massive molecular plug known as a blood clot.

Allosteric regulation. The binding of a small molecule (effector) to a place other than the active site of the enzyme that results in either an increase (positive effector) in enzyme activity or a decrease (negative effector) in enzyme activity.

Effectors. Small molecules that bind to allosteric enzymes that cause either an increase or decrease in enzyme activity associated with a conformational change in the enzyme.

Positive effector. A small molecule that binds to an allosteric enzyme that causes a conformational change in the enzyme making the enzyme more active than before binding.

Negative effector. A small molecule that binds to an allosteric enzyme that causes a conformational change in the enzyme making the enzyme less active than before binding.

Metabolic pathway. The highly regulated sequence of steps by which most of the reactions in the body occur.

Feedback inhibition. A process in which the activation or inhibition of the first reaction in a reaction sequence is controlled by a product of the reaction sequence.

Isozymes. Structurally similar enzymes that come from different tissues of the same organism.

Diagnose. To determine the cause and nature of a disease.

Tissue-type plasminogen activator (TPA). The enzyme administered to heart attack victims that converts the zymogen plasminogen to plasmin, the enzyme that breaks down blood clots.

Chapter 15 Objective Details:

a. An introduction to enzymes. Enzymes are protein catalysts that speed up the chemical reactions that occur in the body at the modest temperatures and mild conditions existing in cells. The names of enzymes end in either the suffix –ase or the suffix –in. The name of the substrate is used as the prefix in the name of the enzyme. A substrate is a reactant in an enzymatic reaction. Enzymes are proteins and like proteins they can be simple or conjugated. The simple protein that consists of only an amino acid chain is referred to as an apoenzyme. If the enzyme contains a prosthetic group, it is called a holoenzyme, which is the equivalent of a conjugated protein. Prosthetic groups can be either metal ions or organic molecules. The metal ions are inorganic substances that are classified as cofactors. Organic molecules like the heme group of hemoglobin or many of the dietary vitamins are both cofactors and coenzymes. The term coenzyme is reserved exclusively for organic molecules where cofactors refer to inorganic or organic substances. The six main classes of enzymes are the oxidoreductases, transferases, hydrolases, lyases, isomerases, and ligases.

b. Enzyme specificity. Enzymes are specific for certain types of substrates. This specificity may be to groups like peptide bonds, or to only D-carbohydrates. For a substrate to be converted to product, it must first bind to a region of an enzyme called the active site. In the lock-and-key model, the active site is thought of as a rigid environment only able to accept a substrate of specific conformation. The induced-fit model is more flexible and forgiving. It states that the binding of substrate to enzyme causes a change in shape that makes the substrate and active site complementary to each other.

c. Enzyme activity. Enzymes can enhance reaction rates by proximity effects, orientation effects, acid-base catalysis, and strain.

d. Rates of enzyme-catalyzed reactions. The rate of an enzyme-catalyzed reaction is dependent upon enzyme concentration, substrate concentration, temperature, pH, and the presence of enzyme inhibitors. Under typical conditions of excess substrate, the rate of an enzymatic reaction will increase linearly with increasing amounts of enzyme. The reverse is not entirely true. As the concentration of substrate increases, at a fixed concentration of enzyme, the reaction rate will increase until the enzyme is saturated. Under saturation conditions, the enzyme is turning over as fast as it can. The maximum rate for the reaction has been achieved and it is represented by the term V_{max}. The rate of an enzyme-catalyzed reaction will increase with increasing temperature until the enzyme begins to denature. The temperature where the reaction is proceeding at its fastest is referred to as the optimal temperature for the reaction. Temperatures higher than or lower than the optimal temperature will have lower reaction rates. Enzyme reactions are also dependent upon pH. Enzymes have an optimal pH value where they function most efficiently. If an enzyme is in an environment that is above or below its optimal pH, the rate of the enzymatic reaction will be decreased.

A group of compounds that affect enzyme activity are the inhibitors. An enzyme inhibitor is a substance that binds to an enzyme and alters its catalytic activity. The two main classes of enzyme inhibitors are the irreversible and reversible inhibitors. Irreversible inhibitors are substances that permanently bind to an enzyme, leaving the enzyme with little or no activity. Reversible inhibitors are substances that bind to an enzyme, reduce the catalytic activity, and then unbind from the enzyme. Among the reversible inhibitors are the competitive inhibitors and the noncompetitive inhibitors. A competitive inhibitor is a substance that resembles the substrate. The inhibitor binds to the active site of the enzyme and prevents substrate binding, thus preventing catalytic activity. Since competitive inhibitors are reversible, they will unbind from the active site allowing the opportunity for substrate to bind. Another class of reversible inhibitor is called the noncompetitive inhibitors. These inhibitors bind to a site on the enzyme other than the active site and cause a conformational change in the enzyme. The conformational change effects or eliminates the ability of the enzyme to catalyze reactions.

e. **Regulation of enzyme activity.** Enzymes can be regulated by the activation of zymogens, reversible covalent modification of the enzyme, and allosteric regulation. Zymogens are proenzymes of protein-cleaving enzymes. A proenzyme is a precursor to an enzyme that has no catalytic activity and a zymogen is a class of proenzyme. Zymogens become catalytic enzymes when one or more peptide bonds are cleaved and a conformation change occurs to form the active enzyme. Zymogens can be recognized by the suffix –ogen or the prefix pro-. Zymogens are typically precursors to digestive enzymes that would destroy their area of synthesis if they were formed as the active enzyme. Reversible covalent modification of an enzyme is another way enzymes can be regulated. Covalent modification involves the covalent attachment of a group to an enzyme that results in the reversible increase or decrease of enzyme activity. Allosteric regulation covers the enzymes that have a binding site other than the active site. When a small molecule called an effector binds to this other site, the activity of the enzyme is either increased in the case of a positive regulator, or decreased in the case of a negative regulator. Allosteric enzymes play a key role in reactions of the metabolic pathway. When you eat, your body must respond by activating digestive enzymes. When your food is digested, your body must respond by no longer producing digestive enzymes. Feedback inhibition is the body's mechanism of shutting down the reactions of the body. A product at the end of the metabolic pathway serves as an inhibitor to the reaction that started the pathway.

Solutions to Odd Numbered Problems:

1. Enzymes are protein catalysts that speed up the chemical reactions that occur in the body at the modest temperatures and mild conditions existing in cells. Enzymes are essential to living organisms because they allow essential chemical reactions to take place that would otherwise destroy their biological environment in the absence of a catalyst.

3. The enzymes are (a) Trypsin, (d) Acetylcholinesterase, (e) catalase, and (f) sucrase.

5. a. Xylulose reductase is the enzyme that catalyzes the reduction of the substrate xylulose.
   b. UDP-glucuronosyl transferase is the enzyme that catalyzes the transfer of atoms between the substrates UDP and glucuronic acid.
   c. Xanthine oxidase is the enzyme that catalyzes the oxidation of the substrate xanthine.

7. a. Phosphohexose isomerase is the name of the enzyme that catalyzes the isomerization of the substrate phosphohexose.
   b. Aspartate-ammonia ligase is the name of the enzyme that forms a bond between the substrates aspartate and ammonia.
   c. Succinyl-CoA hydrolase is the enzyme that catalyzes the hydrolysis of the substrate succinyl-CoA.

9. A substrate is a molecule that reacts in the enzymatic reactions of biological systems.

11. An enzyme that lacks its prosthetic group is an apoenzyme.

13. The active site of an enzyme is the place on the enzyme where the substrate binds and catalysis occurs.

15. A substrate may bind to the active site of an enzyme by hydrogen bonding, hydrophobic interactions, and ionic and other polar interactions.

17. A polypeptide would not bind to fatty acid synthase because the interactions that bind fatty acids to the active site of the enzyme would be missing. Fatty acids have long hydrocarbon chains that are nonpolar where a polypeptide has peptide bonds that are polar.

19. An enzyme or catalyst will lower the activation energy of a reaction.

21. When two molecules of substrate bind to the active site of an enzyme, they are in close proximity to one another. The term used to describe this is the proximity effect. Two substrate molecules brought close to one another in the active site of an enzyme will react more readily than two molecules by chance running into one another.

23. A plot of reaction rate of an enzyme-catalyzed reaction versus enzyme concentration is shown.

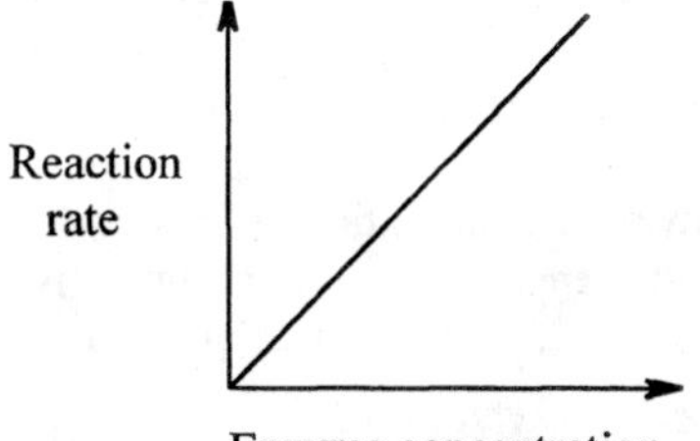

25. When the amount of substrate is in excess, the enzyme can only catalyze the reaction to a maximum velocity. As excess substrate is added, the maximum velocity cannot be changed unless more of the enzyme is added. Without the addition of more enzyme, the increasing amount of substrate causes the reaction rate to level out. If the amount of enzyme is increased, the rate of the reaction will increase because there is more of the enzyme present to catalyze the reaction.

27. Pepsin has an optimal pH of around 2 and trypsin has an optimal pH of around 7. Pepsin is less active to inactive at pH values above or below 2 just as trypsin is less active to inactive at pH values above or below 7. The physiological significance of these pH profiles is that the body produces digestive enzymes that will function in different environments to ensure efficient digestion of food. Pepsin will catalyze hydrolysis reactions in the stomach at low pH while trypsin will catalyze hydrolysis reactions under the far milder conditions of the small intestine where the pH is close to neutral.

29. The dependence of reaction rates due to changes in pH and temperature in the body are very important. If a person's body temperature rises (hyperthermia) due to fever or heat exhaustion, enzyme denaturation can occur resulting in the death of the individual. If a person's body temperature lowers significantly due to cold (hypothermia), the reaction rates for the enzymes of the body will be slowed to the point of being unable to catalyze sufficient reactions to keep the individual alive. Changes in pH are also very important because enzymes only have a narrow range of pH values where they will most efficiently catalyze reactions. Decreases in blood pH due to acidosis can cause enzyme denaturation, and can result in death if not corrected. Alkalosis or high pH levels can also result in enzyme denaturation. Normally, the body maintains narrow ranges of temperature and pH that ensure the activity of enzymes.

31. Both competitive and noncompetitive inhibitors are reversible inhibitors. The difference between the two is that competitive inhibitors mimic the substrate and bind to the active site where noncompetitive inhibitors do not bind the active site, but cause a conformational change in the enzyme by binding to a site other than the active site. In either case, the result is a decrease or elimination of enzyme activity.

33. Sulfa drugs are competitive inhibitors that bind to the active site of the enzyme dihydropteroate synthase. Bacteria need folic acid to grow and the enzyme dihydropteroate synthase produces folic acid from p-aminobenzoic acid. The inhibition of dihydropteroate synthase by sulfa drugs does not allow the substrate (p-aminobenzoic acid) to bind to the active site. Without substrate binding, no folic acid is produced and the bacteria do not grow.

35. A zymogen can be recognized from its name because they either begin with the prefix pro- or end with the suffix –ogen.

37. Trypsin is not synthesized directly in the pancreas because it is a digestive enzyme that would destroy the proteins in the tissues around it. It must be synthesized as the inactive precursor and transported to the small intestine where the tissue is able to tolerate the destructive nature of the enzyme.

39. Glycogen phosphorylase is activated by the binding of phosphate groups.

41. Allosteric refers to enzymes that have more than one binding site. These enzymes have an active site and another site called the allosteric site that will bind molecules called effectors.

43. Positive effectors are small molecules that bind to allosteric enzymes that cause a conformational change in the enzyme making the enzyme more active than before binding. Negative effectors have the opposite effect; they cause a decrease in enzyme activity upon binding to the allosteric site of an enzyme.

45. Feedback inhibition is the process in which the activation or inhibition of the first reaction in a reaction sequence is controlled by a product of the reaction sequence. Feedback inhibition requires a negative effector to work properly. The absence of the negative effector results in the activation of the allosteric enzyme.

47. Lactate dehydrogenase has different forms so that pyruvate can be converted to lactate in the muscles and lactate can be converted to pyruvate in the heart. The different forms of lactate dehydrogenase catalyze the reaction in different directions.

49. Tissue-type plasminogen activator (TPA) is the enzyme administered to heart attack victims that converts the zymogen plasminogen to plasmin, the enzyme that breaks down bloodclots. When an individual has a heart attack, blood clots will often obstruct arteries resulting in oxygen deprivation and cell death. TPA functions to activate the necessary enzymes to break the clot down before tissue damage occurs.

Sample Test:

1. What is the function of enzymes in biological systems?
   a. Structural components of cell membranes.
   b. Provide a source of stored energy.
   c. Catalyze chemical reactions.
   d. Transmission of nerve impulses.

2. How can you identify a substance as an enzyme just by the name?
   a. Enzymes have the suffix -ase.
   b. Enzymes have the suffix -in.
   c. Enzymes have either the suffix –ase or –in.
   d. Enzymes have either the suffix –ase or the prefix pro-.

3. Which of the following pairings between enzyme type and enzyme function is incorrect?
   a. Transferase – catalyzes the transfer of one or more atoms from one substance to another.
   b. Ligase – catalyzes the bonding together of two substrate molecules with the expenditure of energy.
   c. Isomerase – catalyzes the intermolecular rearrangements between different substrates.
   d. Hydrolase – catalyzes hydration or dehydration.

4. Some vitamins are organic molecules that are necessary for enzyme activity. What term describes these vitamins?
   a. Cofactors
   b. Coenzymes
   c. Substrates
   d. Either cofactors or coenzymes.

5. The two components of a conjugated enzyme are
   a. An apoenzyme and a cofactor.
   b. A holoenzyme and a proenzyme.
   c. A coenzyme and a zymogen.
   d. A holoenzyme and a coenzyme.

6. An enzyme active site is the location in an enzyme where
   a. The substrate binds.
   b. A positive regulator binds.
   c. A noncompetitive inhibitor binds.
   d. More than one correct answer.

7. Which of the following enzyme properties does the induced-fit model explain?
   a. An enzyme will bind one and only one substrate.
   b. The reason allosteric enzymes can only bind effectors outside the active site.
   c. The reason product is rapidly released from the active site.
   d. The reason some enzymes will bind substrates that have peptide bonds and no other similarities.

8. How is the rate of an enzymatic reaction affected as a function of increasing temperature with all other factors constant?
   a. The rate of reaction will increase in a linear fashion.
   b. The rate of a reaction will increase to a point and then the rate will remain constant.
   c. The rate of the reaction will increase to a point and then slow down again.
   d. Temperature does not affect the reaction rate.

9. How is the rate of an enzymatic reaction affected as a function of increasing substrate concentration with all other factors constant?
   a. The rate of reaction will increase in a linear fashion.
   b. The rate of a reaction will increase to a point and then the rate will remain constant.
   c. The rate of the reaction will increase to a point and then slow down again.
   d. Substrate concentration does not affect the reaction rate.

10. Which type of inhibitor causes a conformational change in the enzyme resulting in decreased activity?
    a. Competitive inhibitor.
    b. Noncompetitive inhibitor.
    c. Positive effector.
    d. Sulfa drugs.

11. Which of the following substances is not a zymogen?
    a. Chymotrypsinogen b. Fibrin c. Pepsinogen d. Prothrombin

12. Which of the following factors do not effect the rate of enzymatic reactions?
    a. Hypothermia b. Hyperthermia c. Proenzymes d. Reversible inhibitors

13. Which of the following is the equivalent of a noncompetitive inhibitor?
    a. Positive effector
    b. Negative effector
    c. Isoenzyme
    d. Proenzyme

14. Glutamic acid in the active site of isopentenyl diphosphate isomerase aids the conversion of isopentenyl diphosphate to dimethylallyl diphosphate. The role the amino acid, glutamic acid, is associated with which of the following aspects that involve an enzymes active site?
    a. Proximity effects.
    b. Orientation effects.
    c. Acid-base catalysis.
    d. Strain.

15. The enzyme glycogen phosphorylase is responsible for cleaving glucose from glycogen and attaching a phosphate group to the glucose. What classes of enzymes does glycogen phosphorylase fit into?
    a. Transferases and isomerases
    b. Lyases and oxidoreductases
    c. Hydrolases and ligases
    d. Hydrolases and isomerases

16. Biological warfare involves the use of chemical agents such as nerve gas that bind to enzymes like acetlycholineesterase. What kind of inhibitors would make the best biological weapons?
    a. Competitive inhibitors.
    b. Noncompetitive inhibitors.
    c. Reversible inhibitors.
    d. Irreversible inhibitors.

17. If you cut yourself, what enzyme will cause you to stop bleeding?
    a. Prothrombin b. Fibrin c. Fibrinogen d. Tissue-type plasminogen activator

18. Which of the following enzymes has a substrate that is a carbohydrate?
    a. Cellulase
    b. Glutamine synthetase
    c. Pyruvate carboxylase
    d. Aspartate aminotransferase

19. The antibiotic sulfanilamide acts by
    a. Activating zymogens for bacterial enzymes.
    b. Cleaving peptide bonds in bacterial cell walls.
    c. Competetively inhibiting the production of folic acid.
    d. Noncompetetively inhibiting the production of folic acid.

20. In which of the following pairs do the two terms have the same meaning?
    a. Conjugated enzyme and coenzyme
    b. Holoenzyme and proenzyme
    c. Simple enzyme and apoenzyme
    d. Zymogen and isoenzyme

Answers to Sample Test:

1. c
2. c
3. c
4. d
5. a
6. a
7. d
8. c
9. b
10. b

11. b
12. c
13. b
14. c
15. c
16. d
17. b
18. a
19. c
20. c

Chapter 16: Heredity

Chapter Objectives:

1. To recognize the components of nucleic acids.
2. To describe the structure and replication of DNA.
3. To identify the types of RNA and understand their function.
4. To understand how to use the genetic code.
5. To describe protein synthesis in the cell.
6. To understand how DNA can be modified and how genetic variability arises.
7. To explain regulation of gene expression in bacteria.
8. To understand the pros and cons of genetic engineering.

Key Terms: The key terms are listed in the order in which they are encountered in the chapter.

Nucleic Acids. Nucleic acids are polymers of nucleotides.

Nucleotides. Nucleotides, the monomers of nucleic acids, are molecules that contain a nitrogen containing base, a pentose sugar, and one or more phosphate groups.

Nitrogenous base. A molecule containing two or more nitrogen atoms in either a simple or fused ring.

Purine. A bicyclic base with fused five- and six-membered rings. Purine derivatives include adenine and guanine.

Pyrimidine. A monocyclic base with a six-membered ring. Pyrimidine derivatives include cytosine, thymine, and uracil.

Nucleoside. A molecule made up of a sugar and a nitrogenous base.

Deoxyribonucleic acid (DNA). A polymer of deoxyribonucleotides that is found primarily in the nucleus of a cell.

Complementary bases. Specific pairs of bases in nucleic acid structures that hydrogen bond to each other.

Histones. Basic proteins found complexed to eukaryotic DNA.

Chromosome. An individual DNA molecule bound to a group of proteins (histones).

DNA replication. The process by which DNA molecules produce exact duplicates of themselves.

DNA polymerase. The enzyme that forms DNA from deoxyribonucleotides on a DNA template.

Ribonucleic acid (RNA). A polymer made of nucleotide units that plays an essential role in protein synthesis.

Promoter. The site where RNA polymerase binds to begin transcription.

Messenger RNA (mRNA). The form of RNA that specifies the amino acid sequence of proteins.

Ribosome. A small spherical body in a cell on which protein synthesis takes place.

Ribosomal RNA (rRNA). RNA molecules which, together with proteins, make up the ribosomes on which proteins are synthesized.

Transfer RNA (tRNA). RNA molecules that deliver specific individual amino acids to the ribosomes, the sites of protein synthesis.

Transcription. The process by which DNA directs the synthesis of RNA molecules that carry the coded information needed for protein synthesis.

RNA polymerase. The enzyme that catalyzes the process of transcription that gets the code from a DNA template necessary for the synthesis of a specific protein.

Primary transcript. The products of transcription that must be enzymatically altered to produce a functional molecules.

Virus. A tiny particle composed of a core of nucleic acid, a protein coat surrounding the core, and in some viruses, a membranous envelope surrounding the protein coat.

Gene. A portion of a DNA molecule that contains the information necessary to synthesize a particular polypeptide chain.

Codon. A sequence of three nucleotides in mRNA molecules that code for a specific amino acid.

Genetic Code. The code that gives the assignment of the 64 mRNA codons to specific amino acids or start/stop signals.

Translation. The conversion of the code carried by mRNA into an amino acid sequence of a protein.

Anticodon. A series of three nucleotides on tRNA that is complementary to one of the codons in mRNA.

Mutations. Any change that results in the incorrect base sequencing on DNA.

DNA repair enzymes. Enzymes that move along a strand of DNA looking for and replacing incorrect base pairs.

Substitution mutation. A change in the base sequence of DNA that occurs as a result of incorrect "proof-reading". All subsequent DNA molecules will contain the mutation.

Mutagen. A substance or agent that causes a change in the structure of a DNA molecule.

Frame shift mutation. A mutation that shifts the normal reading sequence of bases in DNA by either insertion or deletion of a base or bases.

Induction. The process that turns on the synthesis of proteins when they are needed.

Operator. The site on a strand of DNA where a repressor binds.

Operon. The combination of promoter, operator, and structural genes that regulate protein synthesis.

Repression. The turning off of protein synthesis when specific proteins are not needed.

Structural genes. Genes that contain the information needed to make a protein.

Reverse transcriptase. The enzyme used to make a single strand of DNA that is complementary to a strand of mRNA.

Plasmid. A small circular DNA molecule from a bacterial source.

Restriction endonucleases. Enzymes that cut DNA leaving short single-stranded ends that are then used to connect the DNA pieces to complementary single-stranded ends in plasmids.

Chapter 16 Objective Details:

a. Nucleotides. Nucleic acids are polymers of nucleotides. Both ribonucleic acid and deoxyribonucleic acid are nucleic acids. Ribonucleic acid is the nucleic acid responsible for protein synthesis in cells. Deoxyribonucleic acid contains all of the genetic information for living organisms. Nucleotides are the building blocks of nucleic acids that contain a nitrogen-containing base, a sugar, and one or more phosphate groups. The nitrogen-containing bases are broken down into two categories, the purines and the pyrimidines. The purines, adenine and guanine, contain two heterocyclic rings fused together. The pyrimidines, cytosine, thymine, and uracil, contain a single heterocyclic ring. The nitrogen containing purines and pyrimidines are bonded to a sugar molecule. In 2-deoxyribonucleic acid, the sugar molecule is 2-deoxyribose. In ribonucleic acid, the sugar molecule is ribose. The final component of a nucleotide is the phosphate group that consists of phosphoric acid (H_3PO_4) as it exists at cellular pH values. Nucleosides are nucleotides without the phosphate group.

The nucleotides of DNA and RNA differ in both the sugar and in one of the pyrimidines. DNA contains the sugar 2-deoxyribose and the nitrogenous bases adenine, guanine, cytosine, and thymine. RNA contains the sugar ribose and the nitrogenous bases adenine, guanine, cytosine, and uracil. The symbols for the bases are often used rather than the full word. A diagram is provided to show the similarities and differences between the bases in DNA and RNA.

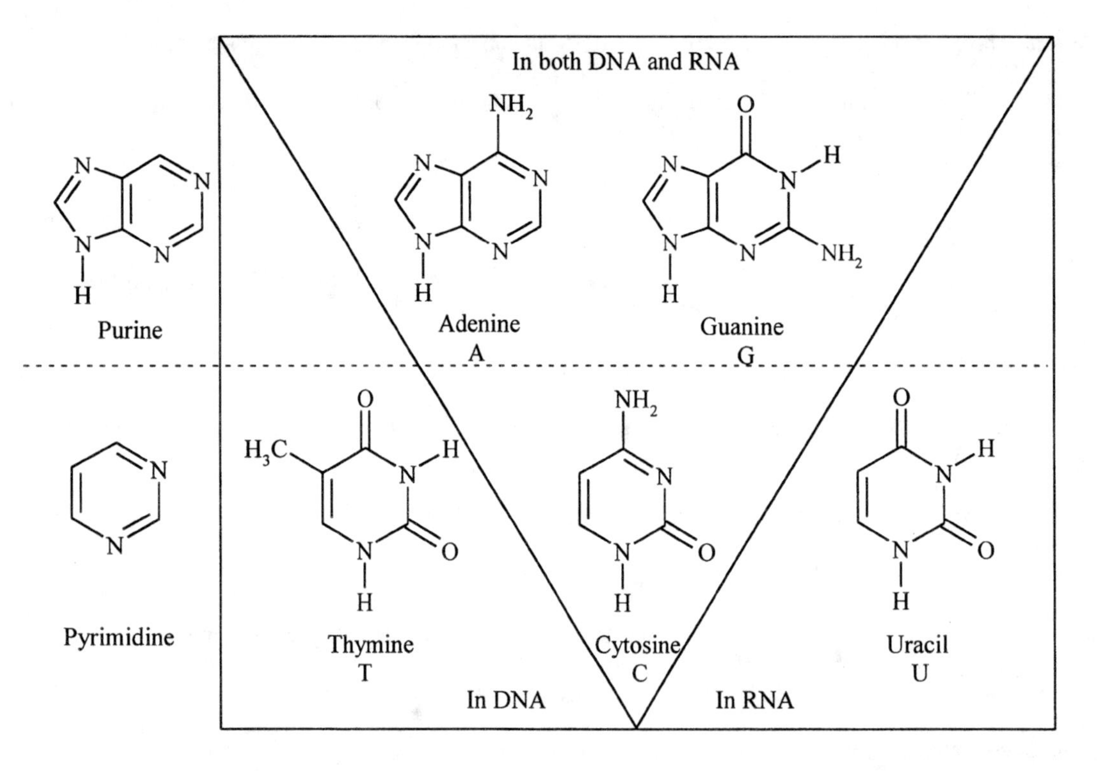

b. DNA. DNA is a macromolecule that is a polymer of deoxyribonucleotides. The nucleotides contain a nitrogenous base, 2-deoxyribose, and a phosphate group. The nitrogenous bases of DNA are A, G, T, and C and they are bonded to carbon number one on the sugar ring. The phosphate group is bonded to the carbon number five on the sugar ring, which leaves the hydroxyl group on carbon number three open for bonding to the phosphate group of another nucleotide. Where proteins are always listed from N-terminal end to C-terminal end, DNA is always listed from 5'-end to 3'-end. The bonding of deoxyribose is shown below.

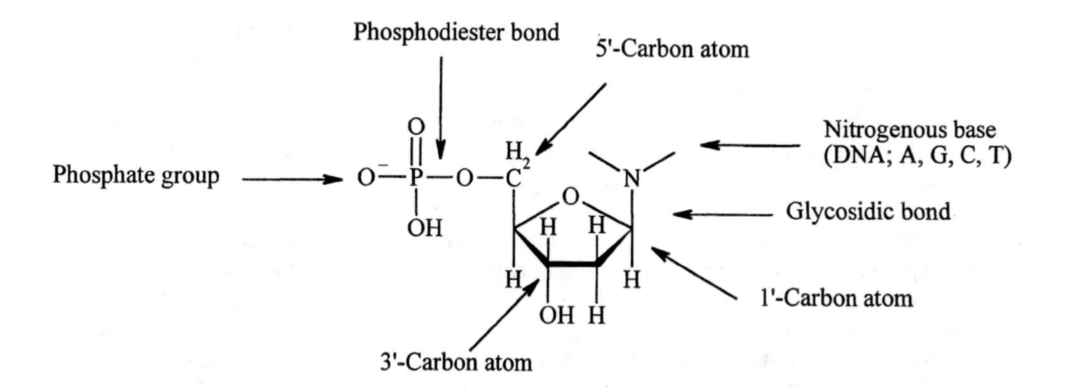

DNA is held in a double helix where the bases of one strand hydrogen bond and share hydrophobic interactions with the bases of a DNA strand traveling in the opposite direction. The base pairs that hydrogen bond to one another are the complementary bases. In DNA, the complementary base pairs are adenine and thymine, and guanine and cytosine. Bacterial DNA is circular while the DNA of higher organisms is linear. The linear DNA is generally found in the nucleus of a cell wràpped around proteins called histones. This DNA/protein complex is known as a nucleosome. When many nucleosomes come together, they contain the genetic information for the cell in a complex called a chromosome.

DNA replication is accomplished by the unwinding of the DNA double helix and the separation of the two DNA strands. The point of separation is called the origin of replication. The enzyme DNA polymerase enters the center of the separated strands and inserts complementary bases that are contained in nucleotides. The bases are matched and the nucleotides are linked together by polynucleotide ligase, which is an enzyme that bonds phosphodiester linkages. The products of DNA replication are two new DNA molecules that are identical to the original DNA molecule. These new strands are known as daughter DNA strands.

c. RNA. RNA is the nucleic acid responsible for making proteins in the cell. RNA differs from DNA because RNA contains the sugar ribose, contains the base uracil in place of thymine, is single stranded, and it is not nearly as large a macromolecule as DNA. RNA functions in transcription and translation. Transcription is the process by which the information contained on a strand of DNA is copied by messenger RNA (mRNA). A section of DNA is unraveled and the enzyme RNA polymerase binds to the DNA template at a site called the promoter. RNA nucleotides are brought in complementary to the DNA template (U instead of T) and a strand of messenger RNA is produced. The mRNA is brought to the ribosomes where ribosomal RNA (rRNA) bound to proteins, called a ribosome, provide a location for protein synthesis. Amino acids are brought to the site of protein synthesis by transfer RNA (tRNA).

d. The genetic code. The genetic code is a list of the 64 possible combinations of the common nitrogenous bases found in RNA that are known as codons. Codons are three-base sequences in mRNA that give rise to specific amino acids or signal start or stop. Messenger RNA is formed in the process of transcription. Transcription involves the formation of a complementary strand of RNA (mRNA) from a region of DNA that codes for a protein known as a gene.

e. Translation (Protein Synthesis). Translation is the process by which cells synthesize proteins. The process begins with the initiation step. Messenger RNA contains the information obtained from DNA. The mRNA strand is complementary to the DNA strand. Initiation begins the process of bringing in codons that are the same as the original gene that was copied by the mRNA. These codons are called anticodons because they are complementary to those of mRNA and the same as those from the original DNA (U has replaced T). Anticodons are found on the bottom of transfer RNA. The anticodon codes for an amino acid. Thus, tRNA brings the amino acids to the ribosomes (site of protein synthesis) according to the message carried by mRNA. The first amino acid in translation is always N-formylmethionine. The fmet- amino acid is a formyl group, a carbonyl group with a hydrogen atom on one side, and the nitrogen of the amino group of a methionine amino acid on the other side. The amino group of the fmet-amino acid is no longer active, but the carboxyl group is still functional. Transfer RNA brings another amino acid into close proximity to the fmet-amino acid and the amino group of the second amino acid forms a peptide bond to the carboxyl group of the fmet-amino acid. A ligase is required for the peptide bond formation. This is part of the process known as elongation. Translocation is the process by which the ribosome moves down to the second amino acid and allows binding of another tRNA molecule that is carrying another amino acid. Transfer RNA continues to bring amino acids to the ribosomes until a stop codon is encountered at which time termination releases the polypeptide.

f. **Genetic variability.** Mutagenesis is the process that leads to permanent changes in the DNA of a cell known as mutation. The two types of mutations are spontaneous and induced. Spontaneous mutations occur when DNA polymerase makes a mistake that isn't caught, or a repair enzyme excises a nucleotide from a parent strand of DNA and inserts the complementary base to an incorrect base. In either case, a permanent change in the DNA has been made that will affect the proteins produced by the transcription and translation of that section of DNA. Induced mutations occur by exposure to mutagens. Various mutagens that cause mutations are electromagnetic radiation including ultraviolet light and x-rays, and chemical agents. Frameshift mutations occur when a base is inserted or deleted from a strand of DNA thus permanently changing the DNA of the cell.

g. **Regulation of gene activity.** Cells regulate the synthesis of proteins by using only some of their genes at any given time. The induction or repression of gene activity allows cell differentiation even though all cells contain the same DNA. The examples you will see are all on bacterial DNA because the DNA is much simpler and easier to work with than eukaryotic DNA. Induction is the process that turns on the synthesis of proteins when they are needed. The region of a gene that codes for a particular regulatory enzyme is called the operon. The operon contains the region of the gene that has the code for the protein (structural gene), a codon that says start here (promoter), and the operator site that binds the repressor protein thus inhibiting RNA polymerase until a substrate is present. When a substrate is present, the repressor protein is removed and RNA polymerase is allowed to make the protein. The making of the protein in the presence of the substrate is known as induction. Repression occurs when an enzyme needed to make a molecule has made enough and must be shut down. In these cases, repression is accomplished by the binding of the enzymatic product to a repressor protein, which then binds to the operator. This binding inhibits RNA polymerase and the enzyme production is terminated. This process is known as repression of gene activity.

h. **Genetic engineering.** Genetic engineering is the process by which genes are added to organisms and proteins are coded for and synthesized by the organism. The initial step is to know what human gene codes for the protein of interest. The gene can be synthesized in the laboratory and inserted into bacterial DNA by the use of a restriction endonuclease followed by a DNA ligase. The bacteria to produce the desired protein can then express the gene that has been inserted into the bacterial DNA.

Solutions to Odd Numbered Problems:

1. The two sugars normally found in nucleic acids are 2-deoxyribose and ribose.

3. The purines are adenine and guanine and the pyrimidines are cytosine, thymidine, and uracil.

5. A nucleoside contains a sugar and a base while a nucleotide contains a sugar, a base, and one or more phosphate groups.

7. The nucleoside containing uracil and ribose.

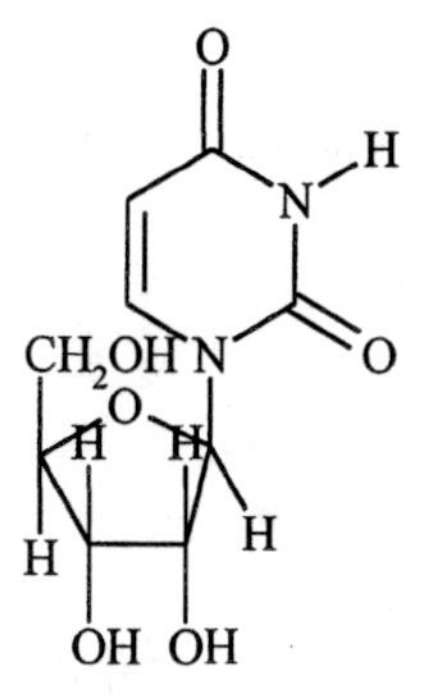

9. Uridine triphosphate

11. Deoxycytidine diphosphate

13. DNA contains the nucleotides deoxyadenosine monophosphate, deoxyguanosine monophosphate, deoxycytidine monophosphate, and deoxythymidine monophosphate.

15. (b) Bases

17. The two strands of DNA are held together by hydrogen bonding interactions and hydrophobic interactions.

19. Adenine and thymine are complementary base pairs that interact through hydrogen bonding and guanine and cytosine are complementary base pairs that interact through hydrogen bonding. For every A in DNA, there must be a T on the complementary strand. The same is true for G and C.

21. a. ACGATG
    b. CTGGTA
    c. GACTGC

23. Pieces of synthesized DNA are connected by the enzyme polynucleotide ligase.

25. The daughter DNA that is produced by replication contains one of the original parent strands of DNA and a newly synthesized strand. Thus the semiconservative nature of the replication is to indicate that half of the new DNA is really DNA that has been conserved from the original DNA.

27. RNA contains the sugar ribose.

29. The four nucleotides found in RNA are adenosine monophosphate, guanosine monophosphate, cytidine monophosphate, and uridine monophosphate.

31. DNA is the largest macromolecule in the cell. It is much larger than RNA.

33. The beginning of transcription on DNA occurs at sites called promoter sites.

35. The answer to this question is not discussed in Section 16.3, but Section 16.6 gives some insight. Transcription occurs over a much smaller area but involves the change from T to U. Replication deals with complementary bases but involves the entire DNA strand. The answer can be justified either way. The explanation given is replication because DNA polymerase possesses "proof-reading capability" not found in RNA polymerase.

37. A gene is a portion of a DNA molecule that contains the information necessary to synthesize a particular polypeptide chain.

39. The DNA base sequence –ACGGTACTG- will have a corresponding mRNA sequence of –UGCCAUGAC-, which will code for the amino acids cysteine, histidine, and aspartic acid.

41. A codon is a sequence of three nucleotides in mRNA molecules that code for specific amino acids and for the start or stop of translation. Sixty-four codons exist in the genetic code.

43. Codons are found in molecules of messenger RNA.

45. The sequence in problem 44 is not the only code for the tripeptide because most amino acids have more than one code for themselves. There are many different codons that will give the same tripeptide. Tryptophan has only one condon.

47. Proteins are synthesized in the ribosomes.

49. Anticodons are found in molecules of transfer RNA.

51. a. AGU
    b. UGC
    c. GCA

53. In the initiation step of translation, a ribosome forms a complex with fmet-tRNA and a mRNA. The fmet-tRNA is the first amino acid in protein synthesis and the ribosome is the location of that synthesis. The formation of the complex is called initiation.

55. Termination of translation occurs when a stop codon on mRNA is encountered by the ribosome. At this point, a ribosomal protein cleaves the polypeptide causing the complex to fall apart and the peptide chain to break off and terminate.

57. (c) Proteins

59. A mutagen is an agent that causes mutation. Some examples of mutagens are ultraviolet light and x-rays.

61. A modification to a base in DNA can result in a different codon. Every three nucleotides contribute to a codon that codes for an amino acid. A change in one base often results in a different codon and consequently a different amino acid. Transcription of DNA by mRNA followed by translation passes on the codon information that is then incorporated into the protein.

63. a. The amino acid serine will be changed to the amino acid threonine.
    b. The amino acid aspartic acid will be changed to the amino acid glutamic acid.
    c. The amino acid threonine will be changed to the amino acid asparagine.

65. A frameshift mutation is a mutation that shifts the normal reading sequence of bases in DNA by either insertion or deletion of a base or bases.

67. Lactose is a compound that turns on a set of genes.

69. An operator is the site on a DNA molecule where RNA polymerase binds in the absence of a repressor. If a repressor is present, the binding of RNA polymerase is inhibited and gene expression does not occur.

71. When a repressor binds to an operator, RNA polymerase no longer has a binding site. In the absence of a binding site, gene expression is shut down. Translation cannot occur when a repressor is bound to an operator.

73. Histidine is a compound that can turn off a set of genes.

75. Restriction endonucleases are used to remove a segment of a plasmid leaving the plasmid with binding sites for the insertion of a synthetic gene.

77. The synthetic gene is not the natural gene, but when transcribed later, it will yield a mRNA that is translated to yield the desired protein. The reason the gene is different is because it must be inserted into the plasmid. For insertion to take place, bases at the ends of the gene must be modified such that they are complementary to the bases left behind by the restriction endonuclease. The core structure of the gene would most likely have a base sequence that resembles the actual base sequence of the gene.

79. There is no such thing as a single unique sequence of bases for the human species. Every individual is different and unique because their DNA is different. The human genome will allow scientists information regarding the big picture of human development. We will be able to more accurately attribute individual human characteristics to individual genes. Some of these characteristics are considered diseases and the knowledge of what bases are responsible for the disease may help in finding a cure.

Sample Test:

1. What are the components of a nucleotide?
   a. One base and one sugar.
   b. Two bases and two sugars.
   c. A base, a sugar, and a phosphate.
   d. Two bases, two sugars, and two phosphates.

2. How many different kinds of nucleotides are present in a single nucleic acid?
   a. Three b. Four c. Five d. There could be billions.

3. Which of the following are pyrimidines?
   a. Adenine, guanine, and uracil.
   b. Adenine, cytosine, and uracil.
   c. Cytosine, thymine, and uracil.
   d. Cytosine, guanine, and thymine.

4. What is the name of the DNA nucleotide that contains the base cytosine, and three phosphate groups?
   a. Cytosine triphosphate
   b. Cytidine triphosphate
   c. Deoxycytosine phosphate
   d. Deoxycytidine triphosphate

5. Where is the free-OH group on the backbone of a DNA molecule?
   a. On the 3'-carbon atom of the ribose sugar.
   b. On the 3'-carbon atom of the deoxyribose sugar.
   c. On the 5'-carbon atom of the ribose sugar.
   d. On the 5'-carbon atom of the deoxyribose sugar.

6. If 15% of the bases in a DNA strand are G, what percent of bases will be A?
   a. 15% b. 30% c. 35% d. 70%

7. Which of the following statements concerning the structure of DNA is incorrect?
   a. DNA is much larger than RNA.
   b. DNA is double stranded while RNA is single stranded.
   c. DNA and RNA both contain the sugar ribose.
   d. DNA contains the base thymine while RNA contains the base uracil.

8. What will be the composition of the DNA strand complementary to –AGCCA-?
   a. –UCGGU-.
   b. –AGCCA-.
   c. –TCGGU-.
   d. –TCGGT-.

9. Which of the following statements concerning the daughter DNA molecules produced during replication is true?
   a. One daughter molecule contains both parent strands and one daughter molecule contains both newly synthesized strands.
   b. Each daughter molecule contains one parent strand and one newly synthesized strand.
   c. Each daughter molecule contains two newly synthesized strands.
   d. Each daughter molecule contains a segment of both parent strands.

10. What RNA sequence will be complementary to the DNA sequence –GTATC-?
    a. –CAUAG-
    b. –CUTUG-
    c. –AUTUG-
    d. –CATAU-

11. What is the name of the region of DNA that carries the information needed for the synthesis of a specific protein?
    a. Codon b. Chromosome c. Gene d. Operator

12. Which of the following are related to codons by the genetic code?
    a. Anticodons b. Amino acids c. Genes d. Base pairs

13. Which of the following types of RNA brings amino acids to the site of protein synthesis?
    a. mRNA b. tRNA c. rRNA d. DNA

14. What is the name of the process by which proteins are synthesized?
    a. Replication
    b. Transcription
    c. Translation
    d. Induction

15. Which of the following is not necessary at the time and place of protein synthesis?
    a. mRNA b. tRNA c. rRNA d. DNA

16. What is the role of restriction endonucleases in genetic engineering?
    a. To serve as a host for foreign DNA.
    b. To cleave the plasmid DNA.
    c. To join the foreign DNA to the host DNA.
    d. To produce proteins in a recombinant plasmid.

17. Which of the following statements about mutations is incorrect?
    a. Mutations can be caused by exposure to ultraviolet light.
    b. Mutations involve a reversible change in the DNA of a cell.
    c. Mutations can occur without apparent outside help.
    d. Mutations can be induced.

18. What word used to describe the process by which DNA directs the synthesis of RNA molecules?
    a. Replication b. Transcription c. Translation d. Expression

19. Which of the following is not part of an operator?
    a. Substrate b. Structural gene c. Operon d. Promoter

20. In what stage of protein synthesis is the amino acid fmet encountered?
    a. Initiation
    b. Elongation
    c. Termination
    d. The amino acid fmet does not exist.

Answers to Sample Test:

1. c
2. b
3. c
4. d
5. b
6. c
7. c
8. d
9. b
10. a
11. c
12. b
13. b
14. c
15. d
16. b
17. b
18. b
19. a
20. a

Chapter 17: Metabolism

Chapter Objectives:

1. To explain the processes of digestion, absorption, and transport.
2. To recognize metabolic pathways and coupled reactions.
3. To explain the role of ATP, NAD^+, and NADPH in metabolism.
4. To describe the origins and fates of the compounds involved in carbohydrate metabolism.
5. To recognize the reactions of the citric acid cycle.
6. To explain how electron transport and oxidative phosphorylation produce ATP.
7. To describe the catabolism and synthesis of lipid molecules.
8. To explain the catabolism of amino acids and the role of the urea cycle.

Key Terms: The key terms are listed in the order in which they are encountered in the chapter.

Catabolism. Biological reactions that produce energy as they break large molecules into smaller ones.

Anabolism. Biological reactions that consume energy as they build larger, more complex molecules from smaller ones.

Digestion. The catabolic reactions of the body that involve the hydrolysis of large food molecules into smaller molecules.

Amylase. The digestive enzyme found in saliva that catalyzes the hydrolysis of glycosidic bonds in starch molecules.

Pepsin. The digestive enzyme found in the stomach that catalyzes the hydrolysis of some peptide bonds in proteins.

Protease. The digestive enzymes found in the small intestine that catalyze the hydrolysis of peptide bonds in proteins.

Peptidases. The digestive enzymes found in the small intestine that catalyze the hydrolysis of peptide bonds in peptides.

Lipases. The digestive enzymes found in the small intestine that catalyze the hydrolysis of ester bonds in triacylglycerols to yield fatty acids, glycerol, and monoacylglycerols.

Lipoproteins. Proteins that possess a hydrophobic interior of lipid coated with a hydrophilic surface of protein molecules that transport lipids throughout the body.

Metabolic pathway. A series of sequential reactions within the body.

Coupled reactions. Reactions of the body that occur concurrently with another reaction.

Adenosine triphosphate (ATP). The nucleotide containing the nitrogenous base adenine, the sugar ribose, and three phosphate groups held together by two high-energy phosphoester bonds that serves as the principal energy-storing compound in cells.

Inorganic phosphate (P_I). The polyatomic ion (PO_4^{3-}) containing one phosphorus atom surrounded by four oxygen atoms with a negative three overall charge.

Nicotinamide adenine dinucleotide (NAD^+). A coenzyme that is reduced and provides energy for catabolic reactions involving oxidation.

Flavin adenine dinucleotide (FAD). A coenzyme that is reduced and provides energy for catabolic reactions involving oxidation.

Nicotinamide adenine dinucleotide phosphate (NADP). A coenzyme that can be reduced and can provide energy for anabolic reactions.

Glycolysis. The principal pathway for catabolism of glucose and other sugars.

Cellular respiration. The aerobic catabolism of pyruvate to acetyl-CoA that is then further broken down to carbon dioxide and water.

Lactate fermentation. The anaerobic catabolism of pyruvate to lactate in a reaction that uses NADH.

Alcohol fermentation. The anaerobic process in yeast that converts pyruvate to acetaldehyde and carbon dioxide and further converts acetaldehyde to ethanol.

Gluconeogenesis. The anabolic synthesis of glucose from lactate, glycogenic amino acids, and glycerol.

Glycogenic amino acids. All of the amino acids with the exception of leucine and lysine.

Glycogenesis. The synthesis of glycogen catalyzed by the enzyme glycogen synthetase.

Heterotrophic. Organisms that must feed on other animals or plants to get the molecules they need.

Autotrophic. Organisms that can take simple inorganic molecules like water and carbon dioxide and use them plus light energy to synthesize organic compounds.

Photosynthesis. The process by which autotrophic organisms synthesize organic compounds.

Pyruvate dehydrogenase. The enzyme that catalyzes the conversion of pyruvate to acetyl-CoA.

Citric acid cycle. A cyclic metabolic pathway that oxidizes acetyl-CoA to carbon dioxide.

Chemiosmotic theory. The theory that explains how the energy stored in NADH and $FADH_2$ is converted to molecules of ATP.

Electron-transport chain. A series of reactions in which electrons and hydrogen ions from NADH and $FADH_2$ are passed to intermediate carriers and ultimately react with molecular oxygen to produce water.

Chemical gradient. A difference in concentration across a semipermeable barrier.

Electrical gradient. A difference in electrical charge across a semipermeable barrier.

Oxidative phosphorylation. The process by which ATP is synthesized from ADP using energy released in the electron transport chain.

Fatty acyl-CoA. The complex that forms along with AMP and pyrophosphate when a fatty acid reacts with coenzyme A in the presence of ATP and water.

Beta oxidation. The oxidation of the beta carbon atom of a fatty acyl group inside the mitochondria to carbon dioxide and water.

Lipogenesis. The process by which fatty acids are synthesized with palmitic acid being the major product.

Fatty acid synthetase. The class of enzymes that catalyze lipogenesis.

Ketone bodies. Water soluble fatty acid derivatives that are commonly found in the blood.

Transamination. The process that removes the amino group from an amino acid and transfers it to α-ketoglutarate.

Oxidative deamination. The process that converts glutamate into α-ketoglutarate and ammonia.

Phenylketonuria. The condition marked by the presence of phenylketones in the urine.

Chapter 17 Objective Details:

a. Digestion, absorption, and transport. Carbohydrate digestion begins in the mouth with the enzyme amylase catalyzing the hydrolysis of glycosidic bonds in starch molecules. The carbohydrates that are not fully digested in the mouth and esophagus are further digested in the small intestine. The enzymes involved in carbohydrate catabolism are amylase, maltase, sucrase, and lactase. Digestion of proteins begins in the stomach with the catabolic enzyme pepsin. Pepsin is able to hydrolyze some peptide bonds, but the majority of protein digestion occurs in the small intestine by the proteases trypsin and chymotrypsin. The small intestine also contains peptidases, which are enzymes that hydrolyze the peptide bonds of oligopeptides. Lipid digestion occurs in the small intestine where lipases catalyze the hydrolysis of ester bonds in triacylglycerols. Lipoproteins are then used to transport the insoluble lipids throughout the body.

b. Metabolic pathways and coupled reactions. Metabolic pathways are a series of sequential reactions within the body. Coupled reactions are reactions of the body that occur concurrently with another reaction.

c. Energy in metabolic reactions. Metabolic reactions are driven by the hydrolysis of ATP to ADP + P_I, the reduction of NAD^+ to NADH, the reduction of FAD to $FADH_2$, or the oxidation of NADPH to $NADP^+$.

d. Carbohydrate metabolism. Glycolysis is the principal pathway for catabolism of glucose and other simple sugars. A summary of key reactions is provided.

Glucose + $2NAD^+$ + 2ADP + $2P_I$ $\rightarrow$ 2 pyruvate + 2NADH + 2ATP + energy
Pyruvate + NADH + H^+ $\leftrightarrow$ lactate + NAD^+ (enzyme = lactate dehydrogenase; anaerobic conditions)
Pyruvate $\leftrightarrow$ Acetaldehyde + CO_2 (enzyme = pyruvate decarboxylase; anaerobic conditions)
Acetaldehyde + NADH + H^+ $\rightarrow$ Ethanol + NAD^+ (enzyme = alcohol dehydrogenase (in yeast); anaerobic conditons.)

Gluconeogenesis is the anabolic synthesis of glucose from lactate, glycogenic amino acids, and glycerol. The net reaction equation is provided.

2Pyruvate + 4ATP + 2GTP + 2NADH $\rightarrow$ glucose + 4ADP + 2GDP + $6P_I$ + $2NAD^+$

Glycogenesis involves the synthesis of glycogen catalyzed by the enzyme glycogen synthetase. Glycogen metabolism involves the addition of a phosphate group and the cleavage of a glucose molecule containing the phosphate. The reactions are summarized below.

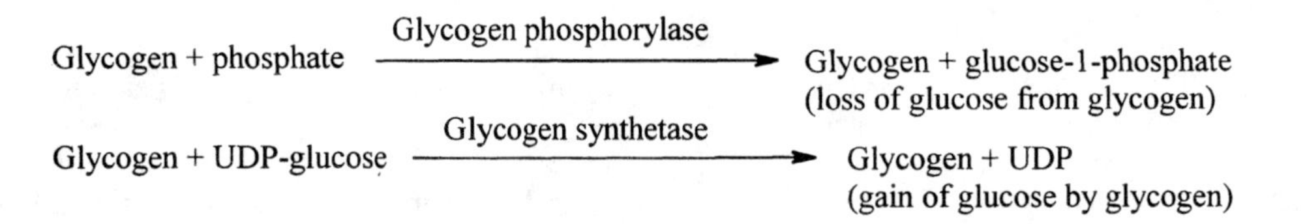

Photosynthesis is the process by which autotrophic organisms convert simple inorganic molecules like water and carbon dioxide and use them plus light energy to synthesize organic compounds. The reaction equation is shown.

$6CO_2 + 6H_2O \rightarrow C_6H_{12}O_6 + 6O_2$ (The reaction requires light.)

e. **The citric acid cycle.** The citric acid cycle is a cyclic metabolic pathway that oxidizes acetyl-CoA to carbon dioxide. Glycolysis breaks glucose down into pyruvate and yields two ATP molecules. Pyruvate contains the energy from glucose so pyruvate dehydrogenase catalyzes the conversion of pyruvate to acetyl-CoA, a molecule that can enter the citric acid cycle. The acetyl-CoA that enters the citric acid cycle transfers energy to the high-energy compounds NADH, $FADH_2$, and GTP that can then be used for further metabolic reactions. The net reaction equation for the citric acid cycle is as follows.

$$\text{Acetyl-CoA} + 3NAD^+ + FAD + GDP + P_i \rightarrow 2CO_2 + CoA + 3NADH + FADH_2 + GTP$$

f. **Electron transport and oxidative phosphorylation.** The electron transport chain is a series of reactions in which electrons and hydrogen ions from NADH and $FADH_2$ are passed to intermediate carriers and ultimately react with molecular oxygen to produce water. Oxidative phosphorylation is the process by which ATP is synthesized from ADP using energy released in the electron transport chain. Every molecule of NADH will be converted to three molecules of ATP and every molecule of $FADH_2$ will be converted to two molecules of ATP. The summation of the past several sections is that glycolysis of glucose produces 6 molecules of ATP, pyruvate dehydrogenase produces another six molecules of ATP, and the citric acid cycle produces 24 molecules of ATP. One molecule of glucose will yield 36 molecules of ATP by the end of the electron transport chain and oxidative phosphorylation.

g. **Lipid metabolism.** Hydrolysis of triacylglycerols releases fatty acids into the blood. The catabolism of fatty acids involves the formation of fatty acetyl-CoA. This is the complex that forms along with AMP and pyrophosphate when a fatty acid reacts with coenzyme A in the presence of ATP and water. Beta oxidation is the oxidation of the beta carbon atom of a fatty acyl group inside the mitochondria to carbon dioxide and water. This process yields a smaller fatty acid, acetyl-CoA, an NADH, and A $FADH_2$. Each of the products of this reaction can then go on to produce ATP in the citric acid cycle, electron transport chain, and by oxidative phosphorylation. Fatty acids can also be synthesized by lipogenesis. Lipogenesis requires carbon dioxide, acety-CoA, and energy. The end result is usually palmitic acid that can be used to store energy as triacyglycerols. Ketone bodies are water-soluble molecules found in the blood that are derived from fatty acids. The three common ketone bodies are acetoacetic acid, beta hydroxybutyric acid, and acetone.

h. **Metabolism of nitrogen-containing compounds.** Transamination is the process that removes the amino group from an amino acid and transfers it to α-ketoglutarate. Catabolism of the alpha-keto acids produces the energy associated with dietary protein. Oxidative deamination is the process that converts glutamate into α-ketoglutarate and ammonia. The ammonia is then reacted with carbon dioxide to produce urea. Nucleotides and nucleic acids are not commonly encountered in the diet, but niacin and riboflavin are required in the diet for proper anabolism of nucleic acids. Both niacin and riboflavin are B vitamins.

Solutions to Odd Numbered Problems:

1. Catabolism breaks big molecules down to smaller molecules and anabolism combines smaller molecules to make bigger ones. Catabolism is used to break down what you feed your body and anabolism is the process by which your body makes what it needs.

3. The role of digestion in animals is to hydrolyze large food molecules into smaller ones.

5. Amylases are produced in the mouth and pancreas and secreted in the mouth and small intestine. More than one amylase is needed because most people don't chew their food sufficiently well to break down starch into glucose. Once amylase enters the stomach, it is denatured, so additional amylase is necessary in the small intestine to finish what the salivary amylase didn't do.

7. The highly acidic environment of the stomach denatures dietary enzymes and exposes their peptide bonds to hydrolysis.

9. Bile salts complex with lipids in the small intestine to form micelles.

11. The absorption of nutrients occurs in the intestines. Intestinal cells absorb the products of digestion.

13.

Food Class	Enzymes	Site of Digestion	Products
Carbohydrates	Amylase, maltase, sucrase, and lactase	Mouth and small intestine	Monosaccharides
Proteins	Proteases and peptidases	Stomach and small intestine	Amino acids
Lipids	Lipases	Small intestine	Fatty acids, monoacylglycerols, and glycerol

15. Two examples of catabolic pathways are the digestion of carbohydrates and the digestion of proteins. Two examples of anabolic pathways are the making of proteins in translation and the making of glycogen in glycogenesis.

17. Many possibilities exist. An example of one of these is the following:
    Isocitrate + $NAD^+ \rightarrow$ α-ketoglutarate + NADH + H^+ + CO_2

19. ATP stands for adenosine triphosphate. ATP is the principle energy-storing compound in cells.

21. NAD^+ and FAD drive catabolic oxidation reactions by becoming reduced to NADH and $FADH_2$, respectively.

23. Enzymes could distinguish between NAD^+/NADH and $NADP^+$/NADPH because the $NADP^+$/NADPH contains a very polar phosphate group that would be exposed to the active site. The presence of the phosphate group could make all the difference in terms of enzyme recognition.

25. a. The first three steps in glycolysis require energy.
    b. The immediate source of energy for the first three steps of glycolysis is ATP.

27. Glycolysis produces two net ATP molecules.

29. Lactate

31. In animals, lactate is produced by fermentation whereas yeast will produce ethanol.

33. Gluconeogenesis is the anabolic synthesis of glucose from lactate, glycogenic amino acids, and glycerol. The role of this pathway in animals is to provide a source of energy and also to allow the synthesis of all other monosaccharides.

35. a. Glycogenic amino acids can be converted to glucose by the process of gluconeogenesis.
    b. All of the amino acids are glycogenic except leucine and lysine.

37. The cleavage of the bond between UDP and glucose provides the energy used to form the glycosidic bond between glucose and glycogen.

39. Glycolysis converts 2% of the energy stored in glucose to useable energy as ATP. The pyruvate that is produced by glycolysis must be converted to acetyl-CoA to be further processed in the citric acid cycle.

41. CO_2 + NADH

43. GTP is produced when Succinyl CoA is converted to Succinate (Step 5).

45. 3 NADH molecules are produced in steps 3, 4, and 8 of the citric acid cycle. $FADH_2$ is produced in step 6 of the citric acid cycle.

47. Oxidation of NADH and $FADH_2$ occurs in the inner membrane of mitochondria.

49. The electrons from the electron transport chain are ultimately accepted by molecular oxygen that gets converted to water in the presence of hydrogen ions.

51. Anytime a substance has a higher concentration in one area than another, it is said to have a chemical gradient. When that substance is charged, as in the case of the hydrogen ions, the gradient is said to be electrical as well.

53. Oxidative phosphorylation refers to the process by which ATP is synthesized from ADP using energy released in the electron transport chain. The electron transport chain generates the energy required to phosphorylate ADP by the process of oxidation.

55. For each molecule of acetyl-CoA that passes through the citric acid cycle, the equivalent of 12 molecules of ATP are generated. For each molecule of glucose that is broken down, two molecules of pyruvate are produced and then converted to two molecules of acetyl-CoA.

57. a. (1 mole Glucose) Glycolysis $\rightarrow$ 6 mol ATP
    b. (2 mol pyruvate) Pyruvate dehydrogenase $\rightarrow$ 6 mol ATP
    c. (2 mol acetyl-CoA) Citric acid cycle and oxidative phosphorylation $\rightarrow$ 24 mol ATP

59. Fatty acids are transported by protein serum albumin in the blood until they are absorbed into the cells. Once inside the cell, the fatty acids are converted to fatty acetyl-CoA. Fatty acetyl-CoA binds to the surface of the mitochondria where the fatty acid is removed from the CoA. The fatty acid binds to a molecule of carnitine for transport across the mitochondrial membrane. Once across the membrane, the fatty acid is liberated and allowed to react with another molecule of coenzyme A.

61. The activation of fatty acids requires the energy from two high-energy bonds that is obtained from the conversion of ATP to AMP.

63. The coenzymes used in beta oxidation are NAD^+ and FAD. The coenzyme used in fatty acid biosynthesis is NADPH.

65. Each double bond in a fatty acid reduces the yield of $FADH_2$ by one, which reduces the ATP yield by two.

67. Palmitic acid is the product of fatty acid synthetase. If other fatty saturated fatty acids are required, an enzyme system can use acetyl-CoA to make longer chain fatty acids. Shorter chain fatty acids can be made by beta oxidation.

69. The three ketone bodies are acetoacetic acid, β-hydroxybutyric acid, and acetone. Ketone bodies are normally used as a source of energy.

71. During a diabetic crisis, ketone bodies can be excessively synthesized causing acidosis. If levels of acetone in the blood reach sufficiently high levels, the smell of acetone can be detected on the breath.

73. Competitive inhibitors of HMG-CoA have helped block the synthesis of cholesterol. The drug lovostatin inhibits the body's ability to synthesize cholesterol.

75. α-ketoglutarate

77. a. Humans excrete nitrogen as urea.
    b. Urea is formed from the reaction of ammonia with carbon dioxide. Ammonia is produced from the oxidative deamination of glutamate.

79. Chemotherapy is designed to inhibit the division of rapidly dividing cells associated with cancer. The human body has many types of cells that divide rapidly as part of their normal growth cycle. Inhibition of rapid cell division also inhibits hair cells, gastrointestinal cells, and blood cells.

81. Other pathways that yield acetyl-CoA include fatty acid catabolism by beta oxidation.

Sample Test:

1. The correct notation for the reduced form of nicotinamide adenine dinucleotide is
   a. NAD^+ b. NAD c. NADH d. $NADH_2$

2. The "fuel" for the citric acid cycle is
   a) acetyl CoA b) citric acid c) citrate ion d) oxaloacetate ion

3. How many molecules of ATP result from the entry of one molecule of $FADH_2$ into the electron transport chain?
   a. 1 b. 2 c. 3 d. 4

4. How many ATP molecules result from the "processing" of one acetyl CoA molecule through the common metabolic pathway?
   a. 2 b. 9 c. 12 d. 36

5. The net yield of ATP from the complete oxidation of one molecule of glucose is
   a. 12 b. 24 c. 36 d. 48

6. The net yield of ATP per glucose molecule during the process of glycolysis is
   a. 2 b. 4 c. 8 d. 12

7. The products in a transamination reaction are
   a. a ketoacid and an amino acid
   b. an amino acid and glycerol
   c. a ketoacid and ammonia
   d. acetyl CoA and ammonium ion

8. Carbon dioxide is a product when lactate is converted to
   a. pyruvate b. ethanol c. acetyl CoA d. ATP

9. Which of the following is a product in oxidative deamination of glutamate?
   a. acetyl-CoA b. water c. pyruvate d. ammonium ion

10. The name of the process in which free glucose is converted to glycogen is
    a. gluconeogenesis
    b. glycogenesis
    c. glycogenolysis
    d. glycolysis

11. In which of the following processes is glucose the end product?
    a. glycogenesis
    b. glyconeogenesis
    c. glycolysis
    d. photosynthesis

12. Which of the following compounds is a *ketone body*?
    a. acetoacetate
    b. oxaloacetate
    c. acetyl CoA
    d. ethanol

13. Which of the following is a possible fate for the acetyl CoA produced from the degradation of a fatty acid?
    a. conversion to pyruvate
    b. ketone body formation

c. cholesterol biosynthesis
d. entrance into the citric acid cycle to produce ATP

14. Amino acid metabolism differs from that of carbohydrates and triacylglycerols in that
    a. there is no storage form for amino acids in the body
    b. amino acids cannot be used for energy production
    c. amino acids cannot be converted to keto acids
    d. amino acids cannot be broken down into small molecules

15. The net effect of transamination is to
    a. collect the amino groups from a variety of amino acids into a single compound
    b. convert all nonessential amino acids into essential amino acids
    c. convert all alpha-amino acids into beta-amino acids
    d. produce energy

16. Where does the majority of digestion occur in the human body?
    a. stomach
    b. pancreas
    c. small intestine
    d. large intestine

17. How many phosphoanhydride bonds does ATP contain?
    a. 1 b. 2 c. 3 d. 4

18. What is the final electron acceptor in the electron transport chain?
    a. H_2 b. $FADH_2$ c. O_2 d. NADH

19. Fatty acid synthesis involves what cofactor?
    a. NADH b. NADPH c. $FADH_2$ d. ATP

20. Beta oxidation of fatty acids yields all of the following products except
    a. acety-CoA b. NADH c. NADPH d. $FADH_2$

Answers to Sample Test:

1. c
2. a
3. b
4. c
5. c
6. a
7. c
8. b
9. d
10. b
11. d
12. a
13. d
14. a
15. a
16. c
17. b
18. c
19. b
20. c